C Programming and Numerical Analysis

An Introduction

Synthesis Lectures on Mechanical Engineering

Synthesis Lectures on Mechanical Engineering series publishes 60–150 page publications pertaining to this diverse discipline of mechanical engineering. The series presents Lectures written for an audience of researchers, industry engineers, undergraduate and graduate students.
Additional Synthesis series will be developed covering key areas within mechanical engineering.

C Programming and Numerical Analysis: An Introduction
Seiichi Nomura
2018

Mathematical Magnetohydrodynamics
Nikolas Xiros
2018

Design Engineering Journey
Ramana M. Pidaparti
2018

Introduction to Kinematics and Dynamics of Machinery
Cho W. S. To
2017

Microcontroller Education: Do it Yourself, Reinvent the Wheel, Code to Learn
Dimosthenis E. Bolanakis
2017

Solving Practical Engineering Mechanics Problems: Statics
Sayavur I. Bakhtiyarov
2017

Unmanned Aircraft Design: A Review of Fundamentals
Mohammad Sadraey
2017

Introduction to Refrigeration and Air Conditioning Systems: Theory and Applications
Allan Kirkpatrick
2017

Resistance Spot Welding: Fundamentals and Applications for the Automotive Industry
Menachem Kimchi and David H. Phillips
2017

MEMS Barometers Toward Vertical Position Detecton: Background Theory, System Prototyping, and Measurement Analysis
Dimosthenis E. Bolanakis
2017

Engineering Finite Element Analysis
Ramana M. Pidaparti
2017

C Programming and Numerical Analysis: An Introduction

Seiichi Nomura

www.morganclaypool.com

ISBN: 9781681733111 paperback
ISBN: 9781681733128 ebook
ISBN: 9781681733135 hardcover

DOI 10.2200/S00835ED1V01Y201802MEC013

A Publication in the Morgan & Claypool Publishers series
SYNTHESIS LECTURES ON MECHANICAL ENGINEERING

Lecture #13
Series ISSN
Print 2573-3168 Electronic 2573-3176

C Programming and Numerical Analysis

An Introduction

Seiichi Nomura
The University of Texas at Arlington

SYNTHESIS LECTURES ON MECHANICAL ENGINEERING #13

MORGAN & CLAYPOOL PUBLISHERS

ABSTRACT

This book is aimed at those in engineering/scientific fields who have never learned programming before but are eager to master the C language quickly so as to immediately apply it to problem solving in numerical analysis. The book skips unnecessary formality but explains all the important aspects of C essential for numerical analysis. Topics covered in numerical analysis include single and simultaneous equations, differential equations, numerical integration, and simulations by random numbers. In the Appendices, quick tutorials for gnuplot, Octave/MATLAB, and FORTRAN for C users are provided.

KEYWORDS

C, numerical analysis, Unix, gcc, differential equations, simultaneous equations, Octave/MATLAB, FORTRAN, gnuplot

Contents

Preface

This book is aimed at those who want to learn the basics of programming quickly with immediate applications to numerical analysis in mind. It is suitable as a textbook for sophomore-level STEM students.

The book has two goals as the title indicates: The first goal is to introduce the concept of computer programming using the C language. The second goal is to apply the programming skill to numerical analysis for problems arising in scientific and engineering fields. No prior knowledge of programming is assumed but it is desirable that the readers have a background in sophomore-level calculus and linear algebra.

C was selected as the computer language of choice in this book. There have been continuous debates as to what programming language should be taught in college. Until around the 1990s, FORTRAN had been the dominating programing language for scientific and engineering computation which was gradually taken over by modern programming languages as PASCAL and C. Today, MATLAB is taught in many universities as a first computer application/language for STEM students. Python is also gaining popularity as a general purpose programming language suitable as the first computer language to be taught.

Despite many options for the availability of various modern computer languages today, adopting C for scientific and engineering computation still has several merits. C contains almost all the concepts and syntax used in the modern computer languages less the paradigm of object-oriented programming (use C++ and Java for that). It has been observed that whoever learns C first can easily acquire other programming languages and applications such as MATLAB quickly. The converse, however, does not hold. C is a compiled language and preferred over interpreted languages for programs that require fast execution.

There is no shortage of good textbooks for the C language and good textbooks for numerical analysis on the market but a proper combination of both seems to be hard to find. This book is not a complete reference for C and numerical analysis. Instead, the book tries to minimize the formality and limits the scope of C to these essential features that are absolutely necessary for numerical analysis. Some features in C that are not relevant to numerical analysis are not covered in this book. C++ is not covered either as the addition of object-oriented programming components offers little benefit for numerical analysis. After finishing this book, the reader should be able to work on many problems in engineering and science by writing their own C programs.

The book consists of two parts. In Part I, the general syntax of the C language is introduced and explained in details. gcc is used as the compiler which is freely available on almost all platforms. As the native platform of gcc is UNIX, a minimum introduction to the UNIX operating system is also presented.

In Part II the major topics from numerical analysis are presented and corresponding C programs are listed and explained. The subjects covered in Part II include solving a single equation, numerical differentiation, numerical integration, solving a set of simultaneous equations, and solving differential equations.

In Appendix A, `gnuplot` which is a visualization application is introduced. The C language itself has no graphical capabilities and requires an external program to visualize the output from the program.

In Appendix B, a brief tutorial of Octave/MATLAB is given. This is meant for those who are familiar with C but need to learn Octave/MATLAB in the shortest possible amount of time.

In Appendix C, a brief tutorial of FORTRAN is given. Again, this is meant for those who are already familiar with C to be able to read programs written in FORTRAN (FORTRAN 77) quickly.

This book is based on the course notes used for sophomore-level students of the Mechanical and Aerospace Engineering major at The University of Texas at Arlington.

Seiichi Nomura
March 2018

Acknowledgments

I want to thank the students who took this course for their valuable feedback. I also want to thank Paul Petralia of Morgan & Claypool Publishers and C.L. Tondo of T&T TechWorks, Inc. for their support and encouragement.

All the programs and tools used in this book are freely available over the internet thanks to the noble vision of the GNU project and the Free Software Foundation (FSF).

Seiichi Nomura
March 2018

PART I

Introduction to C Programming

In Part I, the basic syntax of the C language is introduced so that you can quickly write a program for problems in science and engineering to be discussed in Part II. This is never meant to be a complete reference for the C language. It covers only those items relevant to scientific/engineering computation. However, after Part I, you should be able to explore missing topics on your own. A minimum amount of computer environments is needed and all the programs listed should run on any version of `gcc`.

The only way to learn programming is to write a program by yourself. You never learn programming if you just read books sitting on a sofa.

CHAPTER 1

First Steps to Run a C Program

In this chapter, the basic cycle of running a C program is explained. To execute a C program, it is necessary to first write a C code using a text editor, save the code with the file extension, ".c", launch a C compiler to translate the text into an binary code, and, if everything goes well, run an executable (called `a.out` in UNIX). If this is the first time you program in C, it is important that you try every single step described in the following sections.

1.1 A CYCLE OF C PROGRAMMING

There are a variety of ways to access a C compiler and run a C program. Almost all schools run a UNIX server open to the students. You should be able to activate your account on the UNIX server, connect to the server via an `ssh`[1] client such as PuTTY over the internet, and run a freely distributed C compiler, `gcc`.[2]

It is also possible to have a similar setup at home by running your own Linux server or installing a PC/Mac version of `gcc`. If you come from a Windows or Mac environment, you are accustomed to the graphical user interface (GUI) clicking an icon to open an application. However, to use `gcc`, you must use the character-based interface (CUI) to compile and run your program in a UNIX shell, in a command line (DOS) window (Windows) or in Terminal App (Mac). To run `gcc` on the Windows system, you can go to `www.mingw.org` and download the `gcc` installer, `mingw-get-setup.exe`.

In what follows in this book, we use PuTTY[3] (terminal emulation software) to access a UNIX server and run `gcc` on the server.

Figure 1.1 shows an opening screen when PuTTY is launched on the Windows system. In the box circled, enter the name of a server that runs `gcc` and press the Open button. It will prompt you to enter your username (case sensitive) and password (won't echo back). Once you are logged on the server, you are prompted to enter a command from the console (see Figure 1.2). If you have never used a UNIX system before, you may want to play with some of the essential UNIX commands.

Try the following:

1. Login to the server via PuTTY.

[1] ssh (Secure Shell) is a networking protocol by which two computers are connected via a secure channel.
[2] gcc is an abbreviation for GNU Compiler Collection. It is a compiler system produced by the GNU Project.
[3] PuTTY is a free and open-source terminal emulator available for the Windows system that can be downloaded from `www.putty.org`. The size of the executable is less than 1 MB and the program loads very fast.

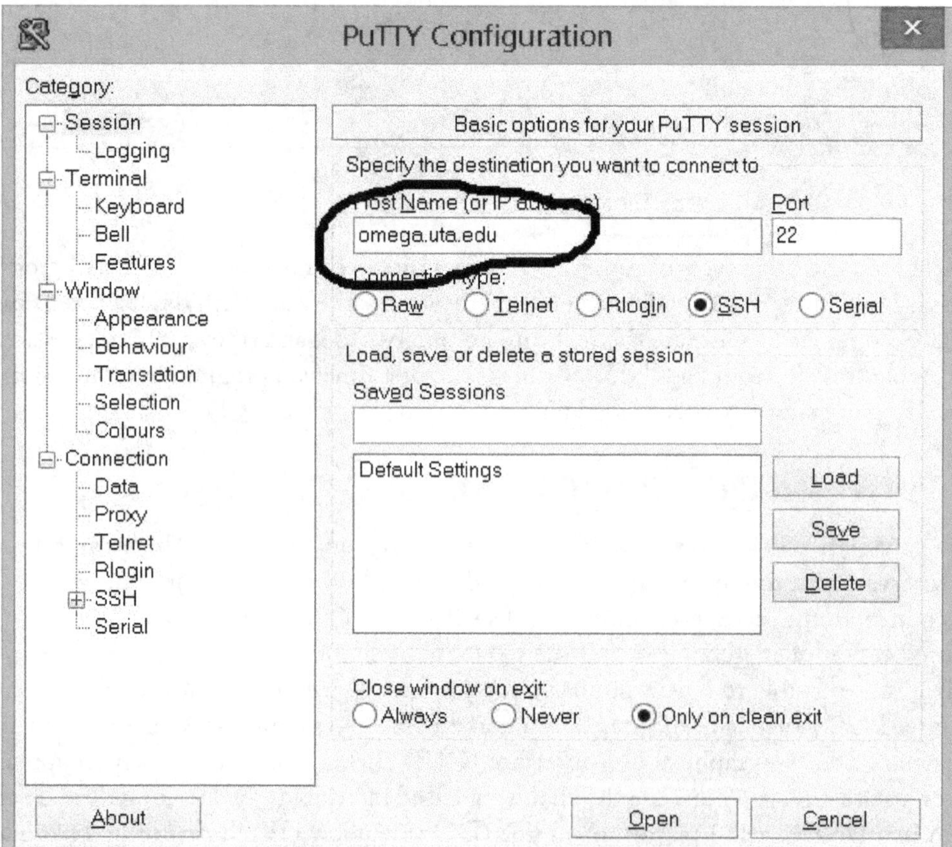

Figure 1.1: Opening screen of PuTTY.

2. Using nano,[4] a simple text editor, compose your C program (Figure 1.3).

```
$ nano MyProgram.c
```

The symbol, $, is the system prompt so do not type it. Enter the following text into nano. Note that all the input in UNIX is case-sensitive.

```
#include <stdio.h>
int main()
{
printf("Hello, World!\n");
return 0;
}
```

[4]nano is a simple editor that comes with all the installation of UNIX. It is a clone of another simple text editor, `pico`.

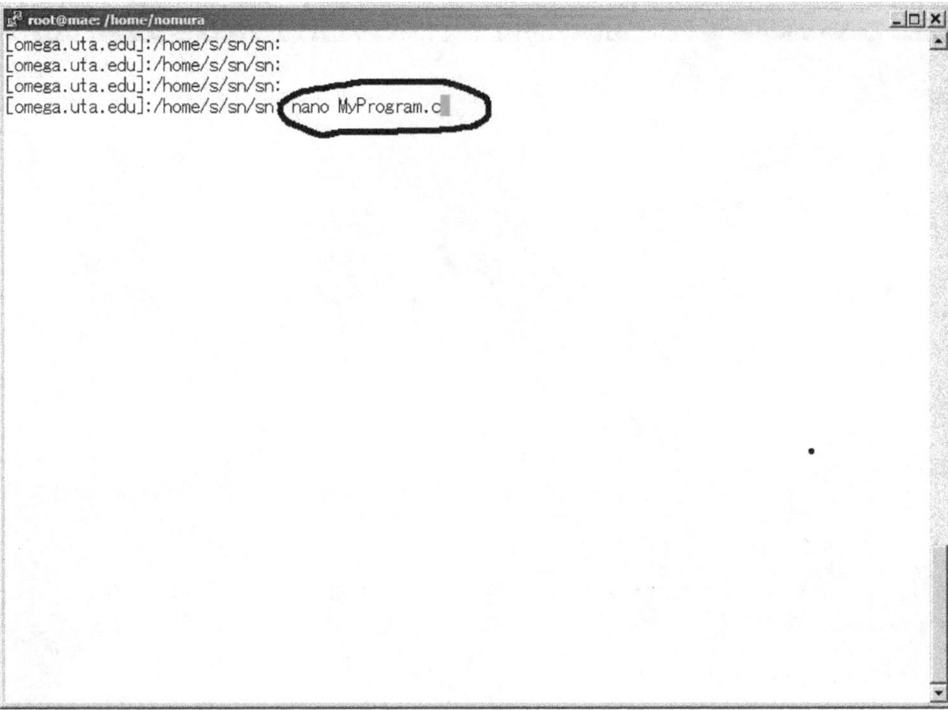

Figure 1.2: A UNIX session in PuTTY.

3. After you finish entering the text, save the file (Control-O[5]) by entering `MyProgram.c`[6] as the file name to be saved and press Control-X to exit from `nano`. This will save the file you just created permanently under the name of `MyProgram.c`.

4. The file you created with `nano` is a text file that is not understood by the computer. It is necessary to translate this text file into a code which can be run on the computer. This translation process is called compiling and the software to do this translation is called a compiler. We use `gcc` for this purpose.

   At the system prompt ($), run a C compiler (`gcc`) to generate an executable file (`a.out`[7]).

   ```
   $ gcc MyProgram.c
   ```

   If everything works, `gcc` will create an executable binary file whose default name is `a.out`.

[5]Hold down the control key and press O.
[6]The file name is case sensitive.
[7]`a.out` is an abbreviation for <u>a</u>ssembler <u>out</u>put.

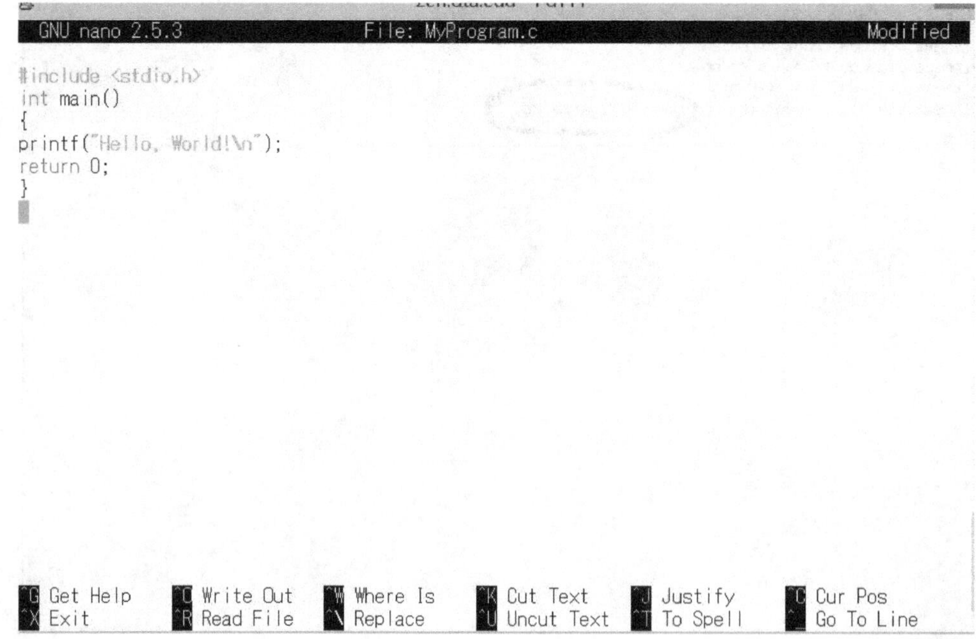

Figure 1.3: A nano session in PuTTY.

5. Run the executable file.

   `$ ./a.out`[8]

6. If there is a syntax error, go back to item 2 and reissue `nano`.

   `$ nano MyProgram.c`

7. If there is no syntax error, run the executable file.

   `$ ./a.out`

8. To logoff from the server, enter `exit`, `logout`, or hit control-D.

1.2 UNIX COMMAND PRIMER

In a perfect world, you could compose a C program, compile it, and run `a.out` and you are done with it. This scenario may work for a program of less than 10 lines but as the size of the program grows or the program depends on other modules, it is necessary to manipulate and organize files on the UNIX system. Even though this is not an introductory book of the UNIX operating

[8]"./" represents the current directory. If the current directory is included in the `PATH` environmental variable, "./" is not necessary.

system, a minimum amount of knowledge about the UNIX operating system is needed. The following are some of the UNIX commands that are used often. Try each command yourself from the system prompt and find out what it does. It won't damage the machine.

- `ls` (Directory listing.)
- `ls -l` (Directory listing in `long` format.)
- `ls -lt | more` (Directory listing, one screen at one time, long format, chronological order.)
- `dir` (alias for `ls`)
- `ls .` (Lists the current directory.)
- `cd ..` (Moves to the directory one level up.)
- `pwd` (Shows the present working directory.)
- `cd /` (Moves to the top directory.)
- `cd` (Returns to the home directory.)
- `mkdir MyNewFolder` (Creates a new directory.)
- `nano myfile.txt` (Creates a new file.)
- `cp program1.c program2.c` (Copies `program1.c` to `program2.c`.)
- `mv old.c new.c` (Renames `old.c` to `new.c`.)
- `rm program.c` (Deletes `program.c`.)
- `rm *.c` (Do not do this. It will delete all the files with extension `c`.)
- `whoami` (Shows your username.)
- `who` (Shows who are logged on.)
- `cal` (Shows this year's calendar.)
- `cal 1980` (Shows the calendar of 1980.)

To quickly move while entering/editing a command line and in `nano` sessions, master the following shortcuts. `^f` means holding down the control key and pressing the `f` key.

- `^f` (Moves cursor forward by one character, `f` for forward.)
- `^b` (Moves cursor backward by one character, `b` for backward.)

- `^d` (Deletes a character on cursor, `d` for delete.)
- `^k` (Deletes entire line, `k` for kill.)
- `^p` (Moves to previous line, same as up arrow, `p` for previous.)
- `^n` (Moves to next line, same as down arrow, `n` for next.)
- `^a` (Moves to top of line, `a` for the first alphabet.)
- `^e` (Moves to end of line, `e` for end.)

1.3 OVERVIEW OF C PROGRAMMING

Arguably, the most important book on the C language is a book known as "K&R" written by Kernighan and Ritchie[9] who themselves developed the C language. It is concise yet well-written and is highly recommended for reading.

1.3.1 PRINCIPLES OF C LANGUAGE

Surprisingly, the C language is based on a few simple principles. They are summarized as follows:

1. A C program is a set of functions.
2. A function in C is a code that follows the syntax below:

```
type name(type var)
{
your C code here.....
......
return value;
}
```

3. A function must be defined before it is used.
4. A function must return a value whose type must be declared (one of `int`, `float`, `double`, `char`). The last line of a function must be a `return xx` statement where `xx` is a value to be returned upon exit.
5. A function must take arguments and must have a placeholder () even if there is no argument.
6. The content of a function must be enclosed by "`{`" and "`}`".

[9]Kernighan and Ritchie, *C Programming Language*, 2nd ed., Prentice Hall, 1988.

7. A special function, `int main()`, is the one which is executed first. It is recommended that this function returns an integer value of 0.

8. All the variables used within a function must be declared.

1.3.2 SKELETON C PROGRAM

The following program is absolutely the smallest C program that can be written:

```
int main()
{
return 0;
}
```

You can compile and execute this program by issuing the following commands:

```
$ gcc MyProgram.c
$ ./a.out
```

where `MyProgram.c` is the name under which the file was saved. Even though it is the smallest C program, the program itself is a full-fledged C code. Of course, this program does nothing and when you issue `./a.out`, the program simply exits after being executed and you are returned to the system prompt.

Here is a line-by-line analysis of the program above. Refer to Section 1.3.1 for the list of items. The first line, `int main()`, indicates that a function whose name is `main` is declared that returns an integer value (`int`) upon exit (Item 4). This function takes no arguments (empty parameters within the parentheses) (Item 5). The program consists of only one function, `main()`, which is executed first (Item 7). The content of the function, `main()`, is the line(s) surrounded by `{` and `}` (Item 6). In this case, the program executes the `return 0` statement and exits back to the operating system returning a 0 value to the operating system. As C is a free-form language, the end of each statement has to be clearly marked. A semicolon `;` is placed at the end of each statement. Hence, `return 0;`.

The following program is a celebrated code that appeared first in the K&R book in the *Getting Started* section and later adapted in just about every introductory book for C as the first C program that prints "Hello, World!" followed by an extra blank line.

```
#include <stdio.h>
int main()
{
printf("Hello, World!\n");
return 0;
}
```

Each line in the above program is now parsed. The first line, `#include <stdio.h>`, is a bit confusing but let's skip this line for the time being and move to the subsequent lines. If you compile the program and execute `a.out`, you will find out that the program prints `Hello, World!` followed by a new line on the screen. Hence, you can guess that the odd characters, `\n`, represents a blank line. As there is no character that represents a blank line, you figure out that `\n` can be used as printing a blank line.

Next, note that the part `printf` is followed by a pair of parentheses and therefore, it is a function in C (Item 5). It is obvious that this function, `printf()`, prints a string, `Hello, World!`, and quits. As it is a function in C, it has to be defined and declared before it is used. However, no such definition is found above the function, `main()`. The first line, `#include <stdio.h>`, is in fact referring to a file that contains the definition of `printf()` that is preloaded before anything else. The file, `stdio.h`, is one of the header (hence the extension, `h`) files available in the C library that is shipped with gcc. As the name indicates (`stdio` = Standard Input and Output), this header file has the definition of many functions that deal with input and output (I/O) functions.

Finally, a function must have information about the type of the value it returns such as `int`, `float`, `double`, etc...(Item 4). In this case, the function `int main()` is declared to return an integer value upon exit. Sure enough, the last statement `return 0;` is to return 0 when the execution is done and 0 is an integer.

Here is how gcc parses this program line by line:

Line 1 Before anything else, let's load a header file, `<stdio.h>`, that contains the definition of all the functions that deal with I/O from the system area.

Line 2 This is the start of a function called `main()`. This function returns an integer value `int` upon exit. This function has no parameters to pass so the content within the parentheses is empty.

Line 3 The `{` character indicates that this is the beginning of the content of the function, `main()`.

Line 4 This line calls the function, `printf()`, that is defined in `<stdio.h>` and prints out a string of `Hello, World!` followed by a blank line. A semicolon, `;`, marks the end of this function.

Line 5 This is the last statement of the function, `main()`. It will return the value 0 to the operating system and exit.

Line 6 The `}` character indicates the end of the content of the function, `main()`.

You can execute this program by

```
$ nano hello.c
```

(Enter the content of the program above.)

```
$ gcc hello.c
```

(If it is not compiled, reedit `hello.c`.)

```
$ ./a.out
Hello, World!
```

Here is another program that does some scientific computation.

```
1:#include <stdio.h>
2:#include <math.h>
3: /* This is a comment */
4:int main()
5: {
6: float x, y;
7: x = 6.28;
8: y=sin(x);
9: printf("Sine of %f is %f.\n", x, y);
10: return 0;
11:}
```

This program computes the value of $\sin x$ where $x = 6.28$. The program can be compiled as

```
$ gcc MyProgram.c -lm
```

Note that the `-lm`[10] option is necessary when including `<math.h>`.[11]

Here is a line by line analysis of the program:

Line 1 The program preloads a header file, `<stdio.h>`.

Line 2 The program also preloads an another header file, `<math.h>`. This header file is necessary whenever mathematical functions such as $\sin(x)$ are used in the program.

Line 3 This entire line is a comment. Anything surrounded by `/*` and `/*` is a comment and is ignored by the compiler.[12]

Line 4 This is the declaration of a function, `main()`, that returns an integer value but with no parameter.

[10]"`-l`" is to load a library and "`m`" stands for the math library.

[11]`<math.h>` only contains protocol declarations for mathematical functions. It is necessary to locally load the mathematical library, `libm.a` by the `-lm` option.

[12]A comment can also start with `//`. This for one-line comment originated in C++.

Line 5 The `{` character indicates that this is the beginning of the content of the function, `main()`.

Line 6 Two variables, `x` and `y`, are declared both of which represent floating numbers.

Line 7 The variable, `x`, is assigned a floating number, 6.28.

Line 8 The function, $\sin(x)$, is evaluated where x is 6.28 and the result is assigned to the variable, `y`.

Line 9 The result is printed. First, a literal string of "`Sine of`" is printed followed by the actual value of `x` and "`is`" is printed followed by the actual value of `y`, a period and a new line.

Line 10 The function, `main()`, exits with a return value of 0.

Line 11 The `}` character indicates that this is the end of the content of the function, `main()`.

There are several new concepts in this program that need to be explained. The second line is to preload yet another header file, `math.h`, as this program computes the sine of a number. In the fourth line, two variables, `x` and `y`, are declared. The `float` part indicates that the two variables represent floating numbers (real numbers with the decimal point). The fifth line says that a number, 6.28, is assigned to the variable, `x`. The equal sign (=) here is not the mathematical equality that you are accustomed to. In C and all other computer languages, an equal sign (=) is exclusively used for substitution, i.e., the value to the right of = is assigned to the variable to the left of =. In the eighth line, the `printf()` function is to print a list of variables (`x` and `y`) with formating specified by the double quotation marks ("..."). The way formating works is that `printf()` prints everything literally within the parentheses except for special codes starting with the percentage sign (%). Here, `%f` represents a floating number which is to be replaced by the actual value of the variable. As there are two `%f`'s, the first `%f` is replaced by the value of `x` and the second `%f` is replaced by the value of `y`. The details of the new concepts shown here will be explained in detail in Chapter 2.

1.4 EXERCISES

It is not necessary to know all the syntax of C to work on the following problems. Each problem has a template that you can modify. Start with the template code, keep modifying the code and understand what each statement does. It is essential that you actually write the code yourself (not copy and paste) and execute it.

1. Write a C program to print three blank lines followed by "Hello, World!". Use the following code as a template:

```
#include <stdio.h>
int main()
{
printf("\nHello, World!\n\n");
return 0;
}
```

\n prints a new line.

2. Write a program to read two real numbers from the keyboard and to print their product. Use the following code as a template. Do not worry about the syntax, just modify one place.

```
#include <stdio.h>
int main()
{
int a, b;  /* to declare that a and b are integer variables */
printf("Enter two integer numbers separated by space =");
scanf("%d %d", &a, &b); /* This is the way to read two integer
 numbers and assign them to a and b. */
printf("The sum of the two numbers is %d.\n", a+b); /* %d is for
        integer format. */
return 0;
}
```

3. Write a program to read a real number, x, and outputs its sine, i.e., $\sin(x)$. You need to use `<math.h>` and the `-lm` compile option. Use the following template program that computes e^x.

```
#include <stdio.h>
#include <math.h>
int main()
{
float x;
printf("Enter a number ="); scanf("%f", &x);
printf("x= %f exp(x)=%f\n",x, exp(x));
return 0;
}
```

You have to use the `-lm` option when compiling:

```
$ gcc MyProgram.c -lm
$ ./a.out
```

CHAPTER 2

Components of C Language

In this chapter, the essential components of the C language are introduced and explained. The syntax covered in this chapter is not exhaustive but after this chapter you should be able to write a simple C program that can solve many problems in engineering and science.

2.1 VARIABLES AND DATA TYPES

Every single variable used in C must have a type which the value of the variable represents. There are four variable types listed in Table 2.1.

Table 2.1: Data types

Type	Content	Format	Range	Example
`int`	Integer	`%d`	–2147483647 ~ +2147483647	10
`float`	Floating number	`%f`	$\pm 2.9387e-39 \sim \pm 1.7014e+38$	3.14
`double`	Double precision	`%lf`	$2^{-63} \sim 2^{+63}$	3.14159265358979
`char`	Character	`%c`	ASCII code	'a'

In Table 2.1, the third column shows the format of each data type which is used in the `print()` and `scanf()` functions.

- `int` represents an integer value. The range of `int` depends on the hardware and the version of the compiler. In most modern systems, `int` represents from -2147483647 to 2147483647.
- `float` represents a floating number. This will take care of most non-scientific floating numbers (single precision). For scientific and engineering computation, `double` must be used.
- `double` is an extension of `float`. This data type can handle a larger floating number at the expense of the amount of memory used (but not much).
- `char` represents a single ASCII character. This data type is actually a subset of `int` in which the range is limited to $0 \sim 255$. The character represented by `char` must be enclosed by a single quotation mark (`'`).

2.1.1 CAST OPERATORS

When an operation between variables of different types is performed, the variables of a lower type are automatically converted to the highest type following this order:

```
int=char < float < double
```

For example, for `a * b` in which `a` is of `int` type and `b` is of `float` type, then, `a` is converted to the `float` type automatically and the result is also of `float` type. There are times when two variables are both `int` type yet the result of the operation is desired to be of `float` type. For example,

```
#include <stdio.h>
int main()
{
  int a, b;
  a=3; b=5;
  printf("%f\n", a/b);
  return 0;
}
```

The output is

```
$ gcc prog.c
2.c: In function 'main':
2.c:6:10: warning: format '%f' expects argument of type 'double',
                   but argument 2 has type 'int' [-Wformat=]
   printf("%f\n", a/b);
          ^
$ ./a.out
-0.000000
```

It prints 0 with a warning even though the result is expected to be 0.6. To carry out this operation as intended,[1] a cast operator (an operator to allow to change the type of a variable to a specified type temporarily) must be used as

```
#include <stdio.h>
int main()
{
  int a,b;
```

[1]Another way of achieving this is to modify `a/b` to `1.0*a/b`.

```
    a=3;b=5;
    printf("%f\n", (float)a/b);
    return 0;
}
```

The output is

```
$ gcc prog.c
$ ./a.out
0.600000
```

The `(float)a/b` part forces both variables to be of `float` type and returns 0.6 as expected.

2.1.2 EXAMPLES OF DATA TYPE

1. This program prints a character, "h".

```
/*
Print a character
*/
#include <stdio.h>
int main()
{
 char a='h';
 printf("%c\n",a);
 return 0;
}
```

   Note that the variable, a, is declared as char and initialized as "h" on the same line.

2. This program prints an integer 10.

```
/*
Print an integer
*/
#include <stdio.h>
int main()
{
 int a=10;
 printf("%d\n",a);
```

```
 return 0;
}
```

Note that the variable, a, is declared as `int` and initialized as `10` on the same line.

3. This program prints a floating number 10.5.

```
/* Print a floating number */
#include <stdio.h>
int main()
{
 float a=10.5;
 printf("%f\n",a);
 return 0;
}
```

Note that the variable, a, is declared as `float` and initialized as `10.5` on the same line.

4. This program prints two floating numbers, 10.0 and -2.3.

```
/* Print floating numbers */
#include <stdio.h>
int main()
{
float a, b=9.0, c;
 a=10.0; c=-2.3;
 printf("a = %f\n",a);
 printf("c = %f\n",c);
 return 0;
}
```

2.2 INPUT/OUTPUT

Almost all C programs have at least one output statement. Otherwise, the program won't output anything on the screen and there is no knowing if the program ran successfully or not. The most common input/output functions are `printf()` and `scanf()` both of which are defined in the header file `stdio.h`.

Use `printf()` (Print with Format) for outputting data to the console and `scanf()` (Scan with Format) for inputting data from the keyboard.

- `printf()`

  The syntax of the `printf()` function is

```
printf("format",argument);
```

  where `format` is the typesetting of the output that you can control and `argument` is a list of variables to be printed. The `printf()` function prints the value(s) of variables in `argument` to the standard output (screen) following the formatting command defined by `format`.

  Examples:

```
printf("Hello, World!\n");
printf("Two integers are %d and %d.\n",a,b);
printf("Two floating numbers are %f and %f.\n",a,b);
printf("Three floating numbers are %f, %f and %f.\n",a,b,c);
```

  A string of characters surrounded by the double quotes (") is printed. However, the percentage sign (%) plus a format letter is automatically replaced by the value of a variable followed. Use `%d` for an integer, `%f` for a floating number, `%lf` for a double precision number, and `%c` for a character. The backslash (\) is called the escape character and escapes the following letter. `\n` represents the next line (inserting a blank line), `\t` represents a tab character and `\a` rings the bell. If you want to print the double quotation mark ("), use `\"`. To print the backslash (\) itself, use `\\`.

- `scanf()`

  The `scanf()` is the inverse of `printf()`, i.e., it scans the value(s) of variable(s) from the standard input (keyboard) with format. The formatting part (i.e., % ...) in `scanf()` is the same as `printf()`. However, the variable name must be preceded by an & (ampersand). The reason why an & is required for `scanf()` but not for `printf()` will be clarified in Section 2.8 (pointers).

  Compare the following two programs:

```
#include <stdio.h>
int main()
{
 int a, b;
```

```
  printf("Enter two integers separated by a comma = ");
  scanf("%d, %d",&a, &b);
  printf("a=%d b=%d\n", a, b);
  return 0;
}
```

The output is

```
$ gcc prog.c
$ ./a.out
Enter two integers separated by a comma = 12, 29
a=12 b=29
```

This program expects that two values are entered from the keyboard separated by a comma (`,`) because of `"%d, %d"` in the `scanf()` function. You have to type the comma (`,`) immediately after the first number. The second number can be entered after as many spaces as you want.

```
#include <stdio.h>
int main()
{
 int a; float b;
 printf("Enter an integer and a real number separated by a space = ");
 scanf("%d  %f",&a, &b);
 printf("a=%d b=%f\n", a, b);
 return 0;
}
```

The output is

```
$ gcc prog.c
$ ./a.out
Enter an integer and a real number separated by a space = 21 6.5
a=21 b=6.500000
```

In this program, two numbers must be entered separated by a space. The number of spaces is arbitrary.

2.3 OPERATORS BETWEEN VARIABLES

There are three types of operators among variables. They are (1) relational operators, (2) logical operators, and (3) increment/decrement operators.

2.3.1 RELATIONAL OPERATORS

The relational operators are mathematical operators often used in the `if` statement. It is important to understand the difference between a single equal sign (=) and double equal signs (==) as the single equal sign (=) is used for an assignment while the double equal signs (==) are used as mathematical equality.

Table 2.2: Relational operators

Symbol	Meaning
<	a < b ; a is less than b.
<=	a <= b ; a is less than or equal to b.
>	a > b ; a is greater than b.
>=	a >= b; a is greater than or equal to b.
==	a == b ; a is equal to b.
!=	a!= b ; a is not equal to b.

Examples

1.
```
if (a==b) printf("a and b are equal.\n");
 else printf("a and b are not equal.\n");
```
   The statement above means that if the two variables, a and `b`, are the same, a string, "`a and b are equal.`", is printed, otherwise a string, "`a and b are not equal.`", is printed. Note the difference between one equal sign (=) and two equal signs (==). One equal sign is for assignment and two equal signs are for logical equality. The double equal signs in C are equivalent to the equal sign in regular mathematical equations.

2.
```
a = 10;
a = a + 1;
```
   The single equal sign in C represents an action to assign the value to the right of the equal sign to the variable to the left of the equal sign. The statement `a = a + 1` does not make sense as a mathematical expression. However, it makes perfect sense as a C statement. `a + 1` is first evaluated to be 11 and this value of 11 is then assigned to `a`. Effectively, the value of `a` was incremented by 1.

2.3.2 LOGICAL OPERATORS

Logical operators in C are mostly used within the `if` statement.

Examples The following examples are self-explanatory for what they do:

Table 2.3: Logical operators

Symbol	Meaning
&&	And
||	Or
!	Not

```
if (a>0 && a<100) printf("a is between 0 and 100.\n");

if (a>0 || a<-5) printf("a is positive or less than -5.\n");

int a=20;
if (!(a==10)) printf("a is not equal to 10.");

if (a==10)
  { printf("The value of a is 10.\n");
    return 0;
  }
else ....
```

2.3.3 INCREMENT/DECREMENT/SUBSTITUTION OPERATORS

A C statement, `a=a+1`, means that the value of a is to be incremented by 1. Five bytes (letters) are used to store `a=a+1`. As C is the language for a minimalist, using 5 bytes for incrementing a variable by 1 seemed to be a waste of memory for the founders of C. Hence, new syntax has been added to shorten this memory allocation. Instead of `a = a + 1`, a++ or ++a may be used.[2] Table 2.4 lists shorthand notations for assignment operations.

Examples

```
i=100;
i++;
```

is the same as

```
i=100;
i=i+1;
```

All of the following statements increment `i` by 1.

```
i=i+1;
```

[2]The naming of C++ follows this, i.e., an "incremental" improvement over the C language.

Table 2.4: Shorthand notations for assignment operations

Symbol	Meaning	
++	$b = {++}a$	a is incremented by 1 and assigned to b. Same as $a = a + 1; b = a;$
	$b = a{++}$	a is assigned to b first and incremented by 1. Same as $b = a; a = a + 1;$
−−	$b = {--}a$	a is decremented by 1 and assigned to b. Same as $a = a - 1; b = a;$
	$b = a{--}$	a is assigned to b first and decremented by 1. Same as $b = a; a = a - 1;$
+=	$a \mathrel{+}= b$	$a + b$ is assigned to a. Same as $a = a + b;$
−=	$a \mathrel{-}= b$	$a - b$ is assigned to a. Same as $a = a - b;$
*=	$a \mathrel{*}= b$	$a * b$ is assigned to a. Same as $a = a * b;$
/=	$a /= b$	a/b is assigned to a. Same as $a = a/b;$
%=	$a \mathrel{\%}= b$	Remainder of a=b is assigned to a. Same as $a = a\%b;$

```
i+=1;
i++;
++i;
```

The difference between `++i` and `i++` is that the former pre-increments `i` before any operation while the latter post-increments `i` after the operation is done.

2.3.4 EXERCISES

1. Write a program that interactively reads temperature in Celsius and convert it to Fahrenheit. Note

$$C = (F - 32) \times \frac{5}{9}.$$

   Expected output:

```
$ ./a.out
Enter temperature in C = 29
It is 84.2 degrees in Fahrenheit.
```

2.4 CONTROL STATEMENTS

Without control statements, all a program can do is to execute the code from the top to the bottom once and for all which is probably not much useful. With control statements, a program can branch out to different parts and repeat operations as many times as need be. The following are the five types of control statements that can control the flow of the program:

- `if .... else`
- `for ( ; ; )`
- `while`
- `do while`
- `switch`

2.4.1 IF STATEMENT

A block (a set of statements surrounded by `{` and `}`) following the `if` statement is executed when the condition inside `(...)` is satisfied. If there is no block, the next statement is executed. `else` is optional.

The following program tells whether an integer entered from the keyboard is between 1 and 100 or not:

```
#include <stdio.h>
int main()
{
 int i;
 printf("Enter an integer = ");
 scanf("%d",&i);
 if (i>1 && i<100)
  printf("The number is between 1 and 100.\n");
 else
 printf("The number is not in that range.\n");
 return 0;
}
```

The output looks like:

```
$ gcc prog.c
$ ./a.out
Enter an integer = 45
The number is between 1 and 100.
$ ./a.out
Enter an integer = 104
The number is not in that range.
```

2.4.2 FOR STATEMENT

A `for` loop is to be used when the number of iterations to be executed is known in advance. The `for` loop consists of three parts each of which is separated by a semicolon (`;`). The first part is initialization of a counter variable, the second part is a test of the iteration counter variable and if this test fails, the loop is finished. The third part defines an action on the counter variable. A block can be followed after the `for` statement to execute multiple statements for each iteration.

```
for (initial value; condition ; counter increment) statement;
```

Examples

1. The following program prints 0–9.

```
#include <stdio.h>
int main()
{
 int i;
 for (i=0; i< 10; i++) printf("i=%d\n",i);
 return 0;
}
```

The output is

```
$ gcc prog.c
$ ./a.out
i=0
i=1
i=2
i=3
i=4
i=5
i=6
i=7
i=8
i=9
```

2. The following program computes

$$S = 1 + 2 + 3 + 4 + 5 + 6 + \ldots + 100.$$

```
#include <stdio.h>
int main()
{
 int i, sum = 0;
 for (i=0; i<= 100; i++) sum = sum  + i;
 printf("Sum = %d\n", sum);
 return 0;
}
```

The output is

```
$ gcc prog.c
$ ./a.out
Sum = 5050
```

In this program, i is a counter variable and sum is a placeholder for the sum. The parameters in the for statement mean that the counter variable, i, is reset to be 0 first and as long as i is less than or equal to 100, the sum = sum + 1 statement is repeated. Upon each execution of sum = sum + 1, i is incremented by 1. In the sum = sum + 1 statement, the previous value of sum is incremented by 1 at each iteration and the newly incremented value of sum+1 replaces the previous value of sum.

The following pattern can compute mathematical summation:

```
sum = 0.0;
for (i=0; i<= 100; i++) sum = sum + f(i);
```

for

$$\sum_{i=0}^{100} f(i) = f(0) + f(1) + f(2) + \ldots + f(100).$$

Note: The iteration variables (i, j, k, ...) must be always declared as int.

3. Approximating ln 2

   It is known that[3]

$$1 - \frac{1}{2} + \frac{1}{3} - \frac{1}{4} + \frac{1}{5} - \ldots = \ln 2. \tag{2.1}$$

[3]Integrating the both sides of the geometric series,

$$\frac{1}{1+x} = 1 - x + x^2 - x^3 + x^4 - x^5 + \ldots$$

Therefore, by numerically summing up the left-hand side of Eq. (2.1), one can obtain a numerical value of ln 2(= 0.693147...). Equation (2.1) can be written as

$$\sum_{i=1}^{\infty} \frac{(-1)^{i+1}}{i} = \ln 2. \tag{2.2}$$

To implement the left-hand side of Eq. (2.2) in C, use the following statement:

```
sum = sum + pow(-1, i+1)/i;
```

Note that `pow(a,b)` is a function found in `<math.h>` which returns a^b. A program to implement Eq. (2.1) should look like:

```
#include <stdio.h>
#include <math.h>
int main()
{
 int i,n;
 float sum=0.0;
 printf("Enter # of iterations = ");
 scanf("%d", &n);
 for (i=1;i<n;i++)
  sum = sum + pow(-1, i+1)/i;
 printf("Approximation of ln(2)= %f. ", sum);
 printf("Exact value of ln(2)= %f.\n", log(2));
 return 0;
}
```

The output is

```
$ gcc prog.c -lm
$ ./a.out
```

one can obtain

$$\ln(1+x) = x - \frac{x^2}{2} + \frac{x^3}{3} - \frac{x^4}{4} + \frac{x^5}{5} - \frac{x^6}{6} + \ldots$$

Substituting $x = 1$ on both sides yields

$$\ln 2 = 1 - \frac{1}{2} + \frac{1}{3} - \frac{1}{4} + \frac{1}{5} - \ldots$$

```
Enter # of iterations = 1000
Approximation of ln(2)= 0.693646. Exact value of ln(2)= 0.693147.
$ ./a.out
Enter # of iterations = 10000
Approximation of ln(2)= 0.693191. Exact value of ln(2)= 0.693147.
$ ./a.out
Enter # of iterations = 10000000
Approximation of ln(2)= 0.693137. Exact value of ln(2)= 0.693147.
```

As is seen from the output above, the convergence of the series is rather slow. Also note that for today's computers, iterations for 10,000,000 times pose no problem whatsoever.

4. We want to find one of the roots of a cubic equation defined by

$$x^3 + x - 1 = 0, \tag{2.3}$$

which is between 0 and 1 by an iterative method. Modify Eq. (2.3) to be read as

$$x = \frac{1}{1 + x^2}. \tag{2.4}$$

Although Eqs. (2.3) and (2.4) are mathematically equivalent, Eq. (2.3) cannot be used as a valid C statement while Eq. (2.4) can be used as a valid C statement where the evaluated value of $\frac{1}{1+x^2}$ is assigned to x. Starting with appropriate initial guess for x, Eq. (2.4) can be iterated until convergence is attained. The following program uses $x = 1$ as initial guess:

```
#include <stdio.h>
int main()
{
 int i,n;
 float x=1.0;
 printf("Enter # of iterations = ");
 scanf("%d", &n);
 for (i=1;i<n;i++) x = 1/(1+x*x);
 printf("Iteration #=%d, x=%f.\n", i, x);
 return 0;
}
```

The output is

```
$ gcc prog.c
$ ./a.out
Enter # of iterations = 30
Iteration #=30, x=0.682327.
$ ./a.out
Enter # of iterations = 31
Iteration #=31, x=0.682328.
$ ./a.out
Enter # of iterations = 32
Iteration #=32, x=0.682328.
```

It is seen that the convergence was attained after 31 iterations. Although the convergence with this iteration method is slow, there is no need to use any advanced mathematics to solve this cubic equation.

2.4.3 `WHILE` STATEMENT

A `while` statement executes the statement(s) that follows while the condition is true.

Note that the following program outputs 10:

```
#include <stdio.h>
int main()
{
 int i=0;
 while (i<10) i++;
 printf("%d\n",i);
 return 0;
}
```

The output is

```
$ gcc prog.c
$ ./a.out
10
```

It appears that the program should output 9, not 10. However, in the test `i<10` when `i=9`, `i` is incremented to 10 and this value is held thereafter.

For multiple statements, it is necessary to use a block with curly brackets (`{` and `}`).

```
#include <stdio.h>
int main()
{
int i=0;
while (i<10)
 {i++;
  printf("i = %d\n",i);
 }
printf("%d\n",i);
return 0;
}
```

The output is

```
$ gcc prog.c
$ ./a.out
i = 1
i = 2
i = 3
i = 4
i = 5
i = 6
i = 7
i = 8
i = 9
i = 10
10
```

2.4.4 DO WHILE STATEMENT

The do while loop is similar to the while loop except that the test occurs at the end of the loop body. This guarantees that the loop is executed at least once before continuing. Such a setup is frequently used where data is to be read. The loop is used to re-read the data if the first set was unacceptable.

The following program keeps prompting until the user enters 0 or 1:

```
#include <stdio.h>
int main()
{
```

```
  int input_value;
do
  { printf("Enter 1 for yes, 0 for no :");
   scanf("%d", &input_value);
  } while (input_value != 1 && input_value != 0);
return 0;
}
```

The output is

```
$ gcc prog.c
$ ./a.out
Enter 1 for yes, 0 for no :10
Enter 1 for yes, 0 for no :2
Enter 1 for yes, 0 for no :-1
Enter 1 for yes, 0 for no :0
```

2.4.5 SWITCH STATEMENT

A `switch` statement can branch out to different tasks depending on the value of the variable used. You can achieve the same result using multiple `if` statements but using `switch` simplifies the program flow.

The following program checks to see if the value entered is either 1 or 2 and prints "`a is neither 1 nor 2.`" if otherwise:

```
#include <stdio.h>
int main()
{
 int i;
 printf("Enter an integer="); scanf("%d", &i);
 switch(i)
 {
  case 1: printf("a is 1\n");break;
  case 2: printf("a is 2\n");break;
  default: printf("a is neither 1 nor 2\n");break;
 }
return 0;
}
```

The output is

```
$ gcc prog.c
$ ./a.out
Enter an integer=12
a is neither 1 nor 2
$ ./a.out
Enter an integer=2
a is 2
```

It is important to use a `break` statement to break out from the block for each statement executed. Note that a colon (`:`) must be used instead of a semicolon (`;`) after `case`.

2.4.6 MISCELLANEOUS REMARKS

There are several remarks in C and UNIX that do not warrant separate sections but are nevertheless worth commenting. They are summarized here.

- Exit from an infinite loop.

  From time to time, the program you wrote may be trapped in an infinite loop and nothing can be done except for closing the window or shutting down the machine. For instance, the following program generates an infinite loop:

```
#include <stdio.h>
int main()
{
int i;
for (i=1; i>0; i++) printf("loop");
return 0;
}
```

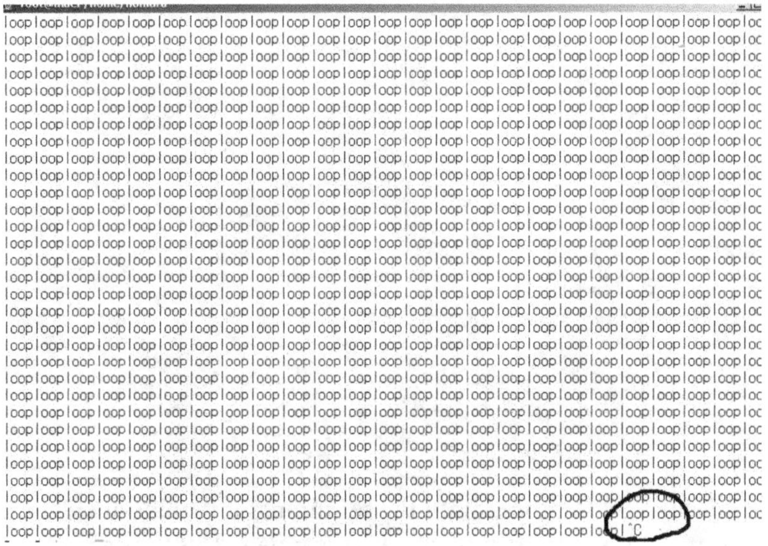

To exit from the infinite loop, enter control-C.[4]

- Output formatting

  It is possible to modify the appearance of the output from the `printf()` function. Study the following format control codes used in the `printf()` statement:

```
#include <stdio.h>
int main()
{
 float a=3.14;
 printf("%f\n",a);
 printf("%10f\n",a);
 printf("%20f\n",a);
 printf("%30f\n",a);

 printf("%10.3f\n",a);
 printf("%10.4f\n",a);
 printf("%10.5f\n",a);
 printf("%10.6f\n",a);
 return 0;
}
```

[4]Hold down the control key and press C. Also if the screen is suddenly frozen and does not accept any keyboard input, try control-Q. This is usually caused by accidentally entering control-S (for pausing output).

The output is

```
$ gcc prog.c
$ ./a.out
3.140000
  3.140000
            3.140000
                      3.140000
     3.140
    3.1400
   3.14000
  3.140000
```

The format, %10.6f means that 10 spaces from the beginning of the line are reserved and the float variable is printed with 6 decimal places right justified. This formatting option is purely cosmetic and does not change the actual value of the variable.

- What is a=b=20 ?

  A statement such as a=b=20 looks strange but it is an acceptable C statement. Try running the following code:

```
#include <stdio.h>
int main()
{
 float a, b;
 printf("%f\n", a=20.0);
 b=a=30.0;
 printf("%d\n", a==20.0);
 return 0;
}
```

The output is

```
$ gcc prog.c
$ ./a.out
20.000000
0
```

In the statement, a=b=20, the control goes toward the end of the line. Hence, b=20 is first executed in which b is assigned 20 but also the statement, b=20, itself is assigned the same value (20) which is subsequently assigned to a. This way, one can assign 20 to both a and b at the same time. In the program above, after b=a=30.0, the value of a is 30. Therefore, a==20.0 is false and hence the printf() function returns 0 (false).

- In gcc, you can use the -o[5] option to specify the name of the executable file instead of a.out.

```
$ gcc MyProgram.c -o MyProgram
```

This generates an executable binary, MyProgram, instead of the default file, a.out, in the same directory. For the Windows version of gcc, the name of the executable is MyProgram.exe.

- You can define a symbolic constant by the define preprocessor:

```
#include <stdio.h>
#define PI 3.141592 /* Defines pi. */
int main()
{
 float a;
 printf("Enter radius = ");
 scanf("%f", &a);
 printf("The area of circle is=%f.\n", a*a*PI);
 return 0;
}
```

The output is

```
$ gcc prog.c
$ ./a.out
Enter radius = 2.0
The area of circle is=12.566368.
```

Whenever the C compiler encounters PI, the compiler replaces PI by 3.141592. It is customary to use upper case letters to define constants with the #define preprocessor.

[5] o for output.

- Why does the `return 0;` in `int main()` have to return 0?

  Returning 0 is not absolutely necessary. In fact, `return -1;` or `return 2019;` works just fine. However, by returning 0 to the operating system when `int main()` exits, the operating system knows that the process finished normally and continues to run accordingly. It does the operating system a favor by telling the operating system that it does not have to bother to do anything special.

2.4.7 EXERCISES

1. Write a C program that interactively reads the three coefficients of a quadratic equation and compute the two roots. The program must alert if there is no real root. The quadratic equation is given by

   $$ax^2 + bx + c = 0,$$

   and the two roots are expressed as

   $$x = \frac{-b \pm \sqrt{b^2 - 4ac}}{2a}.$$

   Note: `sqrt` is available in `<math.h>`. You need to compile the program with the `-lm` option, i.e.,

   ```
   $ gcc MyProgram.c -lm
   ```

2. Write a C program to numerically compute the following series:

   $$1 - \frac{1}{3} + \frac{1}{5} - \frac{1}{7} + \frac{1}{9} - \cdots,$$

   As this series is known to be convergent to $\pi/4$,[6] approximate π using the program. Vary iteration numbers. Note that the general term, a_n, is expressed as

   $$a_n = \frac{(-1)^{n+1}}{2n-1}, \quad n = 1, 2, 3 \ldots$$

[6] Start with the geometric series of

$$\frac{1}{1+x^2} = 1 - x^2 + x^4 - x^6 + x^8 - \ldots$$

Integrate both sides to get

$$\arctan x = x - \frac{x^3}{3} + \frac{x^5}{5} - \frac{x^7}{7} + \frac{x^9}{9} - \ldots$$

Substitute $x = 1$ in the above to get

$$\arctan 1 = \frac{\pi}{4} = 1 - \frac{1}{3} + \frac{1}{5} - \frac{1}{7} + \frac{1}{9} - \ldots$$

2.5 FUNCTIONS

2.5.1 DEFINITION OF FUNCTIONS IN C

The concept of functions in C is important as after all a C program is defined as a set of functions.

You have already seen an example of C function, `main()`, which is always executed first. That is the only special property with `main()`, which means that any function in C must follow the same syntax as `main()`, i.e., declaration of the type (`int` for `main()`), argument(s) within the parentheses (leave it blank if none), all the statements within { and }, and `return value` as the last line in `main()`. If the function does not need any return value, use `void` as the type of return value.

Examples

1. The following program shows an example of the type, `void`. All the user-defined function, `write()`,[7] does is to print "`Hello World!`" and it does not return any value upon exit. Therefore, it has to be declared as `void`.

```
#include <stdio.h>
void write()
{
 printf("Hello, World!\n");
}
int main()
{
 write();
 return 0;
}
```

   The output is

```
$ gcc prog.c
$ ./a.out
Hello, World!
```

   As the function, `write()`, does not have any parameter to pass, it has to be called without any argument as `write()` in the `main` function.

2. The following program defines a function, `cube()`, which returns the cube of a parameter, `x`:

[7]Surprisingly, C does have a built-in function called `write()`.

```
#include <stdio.h>
float cube(float x)
{
 return x*x*x;
}
int main()
{
 float x;
 printf("Enter x = "); scanf("%f",&x);
 printf("The cube of x is %f.\n", cube(x));
 return 0;
}
```

The output is

```
$ gcc prog.c
$ ./a.out
Enter x = 3
The cube of x is 27.000000.
```

Upon exit, `cube(x)` returns x^3 with `x` as the argument passed from the `main` function.

3. The following program computes x^y (x to the power of y)[8] using its own function, `power()`, instead of the `pow()`[9] function available in `<math.h>`:

```
#include <stdio.h>
#include <math.h>
float power(float x, float y)
 {
 return exp(y*log(x));
 }
int main()
{
 float x,y;
 printf("Enter x and y separated by space =");
 scanf("%f %f",&x,&y);
```

[8]If $z = x^y$, taking the natural logarithm of both sides yields $\ln z = \ln x^y = y \ln x$. Hence, $z = e^{y \ln x}$.
[9]`pow(x,y)` in `<math.h>` returns x^y.

```
  if (x<0)
   {
   printf("x must be positive !!\n");
    return 0;
    }
  printf("%f to power of exponent %f is %f.\n", x, y, power(x,y));
  return 0;
 }
```

The output is

```
$ gcc prog.c -lm
$ ./a.out
Enter x and y separated by space =2 4
2.000000 to power of exponent 4.000000 is 16.000000.
$ ./a.out
Enter x and y separated by space =-2 4
x must be positive !!
```

The program exits if x is negative. Otherwise, it computes x^y and prints its value.

2.5.2 LOCALITY OF VARIABLES WITHIN A FUNCTION

Variables used within a function are `local`, i.e., they do not retain the values outside the function. In the following program, the variable name, `sum`, is used for both `f()` and `main()` yet `sum` used within `f()` does not propagate outside the function `f()`.

```
#include <stdio.h>
int f(int n)
 {
  int i,sum=0;
  for (i=1;i<= n;i++) sum=sum+i;
  return sum;
 }
int main()
{
 int i, sum=0;
 for (i=1;i <= 10;i++) sum=sum+i*i;
 printf("%d %d\n", sum, f(10));
```

```
return 0;
}
```

The output is

```
$ gcc prog.c
$ ./a.out
385 55
```

The printed value of sum is the sum defined in main() even though a variable of the same name is returned in the function, f().

2.5.3 RECURSIVITY OF FUNCTIONS

C functions can be used and defined recursively, which means that a C function can call its own within the function.[10] Using the recursive algorithm, a program can be written compactly and efficiently.

Examples

1. Fibonacci numbers

   The Fibonacci numbers, a_n, are defined as

$$a_n = a_{n-1} + a_{n-2}, \quad a_1 = 1, \quad a_2 = 1. \tag{2.5}$$

   With $a_1 = 1, a_2 = 1, a_3$ can be computed as $a_3 = a_2 + a_1 = 1 + 1 = 2$. Similarly, a_4 can be computed as $a_4 = a_3 + a_2 = 2 + 1 = 3$. This way, starting with a_1 and a_2, a_n for any number of n can be computed by repeatedly applying Eq. (2.5).[11]

   As C functions can be defined recursively, coding the Fibonacci number is straightforward as in the program below:

```
#include <stdio.h>
int fibo(int n)
{
 if (n==1) return 1;
 if (n==2) return 1;
 return fibo(n-1) + fibo(n-2);
}
```

[10]In old languages such as FORTRAN, recursivity is not supported.

[11]There is an interesting background story with this number. See www.maths.surrey.ac.uk/hosted-sites/R.Knott/Fibonacci/fibnat.html#Rabbits.

```
int main()
{
 int i;
 printf("Enter n = "); scanf("%d", &i);
 printf("%d\n", fibo(i));
 return 0;
}
```

The output looks like

```
$ gcc prog.c
$ ./a.out
Enter n = 12
144
```

The program simply copies Eq. (2.5) to the definition of `fibo()`.[12]

2. Compute $1 + 2 + 3 + 4 + \ldots + n$ using the recursive algorithm.

   If we define

$$\texttt{sum}(n) \equiv 1 + 2 + 3 + 4 + \ldots + n,$$

   the following relation holds:

$$\texttt{sum}(n) = \texttt{sum}(n-1) + n, \quad \texttt{sum}(0) = 0.$$

   Using this recursive property, the following program can compute

$$1 + 2 + 3 + \ldots + n$$

   without using the `for` statement.

```
#include <stdio.h>
int sum_of_integers(int n)
{
 if (n==0) return 0;
 return n + sum_of_integers(n-1);
```

[12]The explicit formula of the Fibonacci sequence is given by

$$a_n = \frac{\left(\frac{1}{2}\left(1+\sqrt{5}\right)\right)^n - \left(\frac{1}{2}\left(1-\sqrt{5}\right)\right)^n}{\sqrt{5}}.$$

```
}
int main()
{
 int n;
 printf("Enter n = ");
 scanf("%d", &n);
 printf("1+2+....+%d =%d \n", n, sum_of_integers(n));
 return 0;
}
```

The output is

```
$ gcc prog.c
$ ./a.out
Enter n = 100
1+2+....+100 =5050
```

2.5.4 RANDOM NUMBERS, `RAND()`

Random numbers can be generated by using `rand()` in `<stdlib.h>` to perform numerical simulations of various experiments that are hard to carry out otherwise. The following program prints random numbers 10 times:

```
#include <stdio.h>
#include <stdlib.h>
int main()
{
 int i;
 for (i=0; i < 10; i++)  printf("%d\n", rand());
 printf("\nMAX = %d\n", RAND_MAX);
 return 0;
}
```

The output is

```
$ gcc prog.c
$ ./a.out
1804289383
846930886
```

```
1681692777
1714636915
1957747793
424238335
719885386
1649760492
596516649
1189641421

MAX = 2147483647
```

The function, `rand()`, returns an integer between 0 and `RAND_MAX` (system dependent)[13] which is defined in the header file, `<stdlib.h>`.[14]

If floating random numbers between 0 and 1.0 are desired instead of between 0 and `RAND_MAX`, the following program is used:

```
#include <stdio.h>
#include <stdlib.h>
int main()
{
 int i;
 for (i=0; i< 10; i++)  printf("%f\n", 1.0*rand()/RAND_MAX);
 return 0;
}
```

The output is

```
$ gcc prog.c
$ ./a.out
0.840188
0.394383
0.783099
0.798440
0.911647
0.197551
0.335223
0.768230
```

[13]2147483647 for most systems. It is the maximum integer value handled by the system.
[14]`stdlib` = standard library.

```
0.277775
0.553970
```

Note the `%f` format and the factor of 1.0. The value of `rand()/RAND_MAX` alone returns 0 as both `rand()` and `RAND_MAX` are integers and the result is a truncated integer.

However, the same numbers are output again and again each time the program is run. They are all predictive and not random numbers. In order to generate a different sequence of random numbers each time the program is run, `srand()`[15] available in `<stdlib.h>` must be used in conjunction with `rand()`. The `srand()` function sets its argument as the seed for a new sequence of pseudo-random integers to be returned by `rand()`. These sequences are repeatable by calling `srand()` with the same seed value. If no seed value is provided, the `rand()` function is automatically seeded with a value of 1.

```
#include <stdio.h>
#include <stdlib.h>
int main()
{
 int i;
 printf("Enter seed integer = ");
 scanf("%d", &i);
 srand(i);
 printf("%d\n", rand());
 return 0;
}
```

The output looks like

```
$ gcc prog.c
$ ./a.out
Enter seed integer = 10
1215069295
```

In the program above, `srand(i)` takes a seed number `i` and generates a random number based on the value of `i`. The problem with this approach is that if the same seed number is given, the `rand()` function returns the same value.

Using the `time()` function defined in `<time.h>`, you can generate a different seed number every time. The function, `time()`, with the argument `NULL` returns the elapsed time since 00:00:00 GMT, January 1, 1970,[16] measured in seconds.

[15] `srand` = seed random.
[16] This is the birthday of UNIX!

```
#include <stdio.h>
#include <time.h>
int main()
{
 int i;
 printf("%d\n", time(NULL));
 return 0;
}
```

The output is

```
$ gcc prog.c
$ ./a.out
1344034011
```

At the time of writing, 1,344,034,011 seconds have passed since 1/1/1960. As this value keeps increasing, it can be used as a seed number for `srand()`. The following program generates a different random number every time it is called.[17]

```
#include <stdio.h>
#include <stdlib.h>
#include <time.h>
int main()
{
 srand(time(NULL));
 printf("%d\n", rand());
 return 0;
}
```

The output may look like

```
$ gcc prog.c
$ ./a.out
1819029242
(Wait a few seconds.)
$ ./a.out
1672748932
```

[17]If the program is run twice within 1 second, the same random number is generated.

Generating Random Numbers in Various Ranges

With an integer number generated by `rand()` and `MAX_RAND`, it is possible to convert this number to any range, be it an integer range or a float number range. The following list shows examples of various ranges generated by `rand()`:

1. `rand()` returns an integer between 0 and `RAND_MAX`.
2. `1.0*rand()/RAND_MAX` returns a floating number between 0 and 1.
3. `5.0*rand()/RAND_MAX` returns a floating number between 0 and 5.
4. `10.0*rand()/RAND_MAX-5` returns a floating number between −5 and 5.
5. `rand()%7`[18] returns an integer of 0, 1, 2, 3, 4, 5, or 6.
6. `rand()%7+10` returns an integer of 10, 11, 12, 13, 14, 15, or 16.

Using Random Numbers to do Simulations (Monte Carlo Method)

Using random numbers to conduct numerical simulations is called the Monte Carlo method.[19] As an example, consider the following integral:

$$\int_0^1 \sqrt{1-x^2}dx = \frac{\pi}{4}.$$

This integration represents the shaded area in Figure 2.1. As it is a quarter of the whole unit

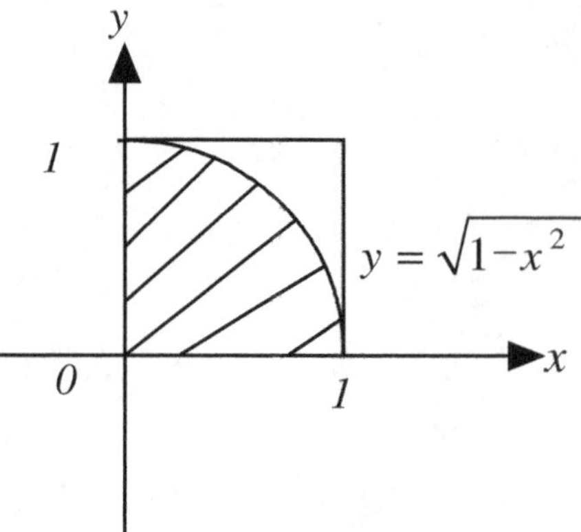

Figure 2.1: Monte Carlo method for numerical integration.

[18] `a%b` returns the remainder of `a/b`.
[19] Monte Carlo is a city in Monaco where one of the major businesses is casinos.

circle, its area is $\pi/4$.[20]

Instead of carrying out the integral directly, the Monte Carlo method can be used to approximate this area. Hence, an approximate value of π can be obtained. To see how this works, a pair of two random numbers, x and y, whose range is between 0 and 1, can be identified as a point, (x, y), inside the square defined by $\{(x, y), 0 \leq x \leq 1, 0 \leq y \leq 1\}$ in Figure 2.1. If (x, y) satisfies $x^2 + y^2 < 1$, the point, (x, y), is inside the shaded area. If not, (x, y) is outside the shaded area. Now, generate a pair of two random numbers between 0 and 1 for N times. For each round, if $x^2 + y^2 < 1$ is satisfied, increment the counter i by 1. If not, go to the next round. At the end of N rounds, the value of i/N should be proportional to the shaded area/the square. As N increases, it is expected that the ratio of i/N is close to $\frac{\pi}{4}/1$. This can be implemented by the following C program:

```
#include <stdio.h>
#include <math.h>
#include <stdlib.h>
#include <time.h>
#define PI 3.141592
int main()
{
 float x, y;
 int i, count=0;
 int n;
 printf("Enter iteration number = ");scanf("%d", &n);
 srand(time(NULL));
 for (i=0; i< n; i++)
 {
  x=1.0*rand()/RAND_MAX;
  y=1.0*rand()/RAND_MAX;
  if (x*x+y*y < 1.0) count=count+1;
 }
 printf("True value = %f\n", PI/4);
 printf("Appx value = %f\n", 1.0*count/n);
 return 0;
}
```

The output looks like

[20]Mathematically, one can integrate the function directly as

$$\int_0^1 \sqrt{1-x^2}dx = \left[\frac{1}{2}\left(\sqrt{1-x^2}\,x + \sin^{-1}(x)\right)\right]_0^1 = \frac{\pi}{4}.$$

```
$ gcc prog.c -lm
$ ./a.out
Enter iteration number = 100
True value = 0.785398
Appx value = 0.790000
$ ./a.out
Enter iteration number = 1000
True value = 0.785398
Appx value = 0.814000
$ ./a.out
Enter iteration number = 100000
True value = 0.785398
Appx value = 0.787400
```

As is seen in the output above, the convergence by the Monte Carlo method is at best mediocre and should be used only as the last resort for 1D integrals. However, the Monte Carlo method is a quick way to approximate 2D and higher-dimensional integrals.

2.5.5 EXERCISES

1. A sequence a_n is given with the following rule:

$$a_{n+2} = -2a_{n+1} + 3a_n, \quad a_0 = 2, \quad a_1 = -1.$$

   Write a C program to compute a_{17}.

2. (a) Write a function, `int factorial(int n)`, which returns $n!$ (the factorial of n, i.e., $1 \times 2 \times 3 \times \ldots \times n$) using the recursive algorithm.

   (b) Using `int factorial(int n)` above, write a program to compute

$$1 + \frac{1}{1!} + \frac{1}{2!} + \ldots + \frac{1}{11!}.$$

3. Using the Monte Carlo method, integrate

$$\int_0^1 \frac{1}{1+x^2} dx,$$

   numerically. Vary the number of iterations (10, 100, 1,000) and estimate an appropriate number of iterations to obtain good accuracy. Note: The exact value of the above integration is $\pi/4$ so this can be also used to approximate the value of π.

4. Using the Monte Carlo method, estimate the volume of the unit sphere

$$x^2 + y^2 + z^2 \leq 1.$$

5. Using the Taylor series for $\cos(x)$ expressed as

$$\cos(x) = 1 - \frac{x^2}{2!} + \frac{x^4}{4!} - \frac{x^6}{6!} + \dots \tag{2.6}$$

create your own $\cos(x)$ and demonstrate your program by making a table which may look like the one shown in Table 2.5, where `mycos(x)` is from your program and $\cos(x)$ is the math function defined in `<math.h>`.

Table 2.5: $\cos(x)$ table

x	mycos(x)	cox(x)
0.0	1.000	1.000
0.1	1.101	1.105
0.2	1.308	1.221
...	...	...
1.0	1.69	1.781

Note:

(a) The values above are false.

(b) Equation (2.6) can be written as

$$\cos(x) = \sum_{i=0}^{\infty} \frac{(-1)^i\, x^{2i}}{(2i)!}.$$

(c) Take the first 10 terms for the series.

Template:

```
#include <stdio.h>
#include <math.h>
int factorial(int n)
{
(your code here)
}
float mycos(float x)
{
 float sum=0; int i;
 for (i=0;i<=10;i++) sum = sum + (your code here);
```

```
  return sum;
 }
 int main()
 {
  float ...;
  int i=1; /* counter */

  (for i=1;i< .......) printf.......
 return 0;
 }
```

2.6 ARRAYS

Vectors and matrices in linear algebra can be implemented in C as arrays. An array in C can represent many elements by a single variable name. This section explains the basic concept of arrays. Arrays are closely related to pointers in C which will be further explained in Section 2.8.

2.6.1 DEFINITION OF ARRAYS

An array is a variable which can represent multiple elements such as numbers and characters. An array can be declared just like other variables as

```
#include <stdio.h>
int main()
{
 float a[3];
 a[0]=1.0; a[1]=2.0; a[2]=5.0;
 (.......)
 return 0;
}
```

The program above defines an array, a, which represents three elements. The values, 1.0, 2.0, and 5.0, are assigned to the first element, the second element and the third element of a, respectively. Note that the index in arrays begins with 0 rather than 1 and ends with $n-1$ where n is the number of elements. This poses a little confusion when arrays are used to represent matrices or vectors in linear algebra as the index in linear algebra begins at 1 so care must be taken when manipulating vector/matrix indices. You can initialize arrays at the same time they are declared as

```
#include <stdio.h>
int main()
{
 float a[3]={1.0, 2.0, 3.0};
 (......)
 return 0;
}
```

or simply,

```
#include <stdio.h>
int main()
{
 float a[]={1.0, 2.0, 3.0};
 (......)
 return 0;
}
```

i.e., if an array is initialized, its dimension can be omitted as the number of elements is automatically determined.

The following program computes the sum of all the numbers in an array:

```
#include <stdio.h>
#define N 5
int main()
{
  int i;
  float a[N]={2.0, -15.0, 12.0, -5.4, 1.9};
  float sum=0.0;
  for (i=0;i <N;i++) sum = sum + a[i];
  printf("The sum is = %f.\n", sum);
  return 0;
}
```

The output is

```
$ gcc prog.c
$ ./a.out
The sum is = -4.500000.
```

2.6.2 MULTI-DIMENSIONAL ARRAYS

Arrays can be nested, i.e., they can take more than 1 indices. Nested arrays (multi-dimensional arrays) can represent matrices in linear algebra. For example, the components of a 2×5 matrix, a, can be represented in C as

$$\mathrm{a} = \begin{pmatrix} a[0][0] & a[0][1] & a[0][2] & a[0][3] & a[0][4] \\ a[1][0] & a[1][1] & a[1][2] & a[1][3] & a[1][4] \end{pmatrix}.$$

Note that the index begins with 0, not 1.

The following program defines a 2×5 matrix, `mat`, given as

$$\mathrm{mat} = \begin{pmatrix} 1.0, & 2.0, & 3.0, & 4.0, & 5.0 \\ 6.0, & 7.0, & 8.0, & 9.0, & 10.0 \end{pmatrix},$$

and prints all the elements using double indices, `i` and `j`.

```
#include <stdio.h>
#define COL 5
#define ROW 2
int main()
{
  int i,j;
  float mat[ROW][COL]={{1.0 ,2.0 ,3.0, 4.0 ,5.0},{6.0, 7.0,
                    8.0, 9.0, 10.0}};
  for (i=0;i<ROW;i++)
   {for (j=0;j<COL;j++)
   printf("%f ",  mat[i][j]); printf("\n");}
  return 0;
}
```

The output is

```
$ gcc prog.c
$ ./a.out
1.000000 2.000000 3.000000 4.000000 5.000000
6.000000 7.000000 8.000000 9.000000 10.000000
```

2.6.3 EXAMPLES

1. Standard deviation and variance

The average of a sequence, $\{x_1, x_2, x_3, \ldots x_n\}$, is the arithmetic average of the components defined as

$$\bar{X} \equiv \frac{1}{N}\sum_{i=1}^{N} x_i.$$

The variance of the same sequence is defined as

$$s_x^2 \equiv \frac{1}{N-1}\sum_{i=1}^{N}(x_i - \bar{X})^2. \tag{2.7}$$

Note the factor, $N-1$, instead of N in Eq. (2.7). This is a mathematical necessity. The standard deviation, s_x, having the dimension of x_i, is defined as

$$s_x \equiv \sqrt{\frac{1}{N-1}\sum_{i=1}^{N}(x_i - \bar{X})^2}.$$

The following program computes the average and the standard deviation of 10 data points:

```
#include <stdio.h>
#include <math.h>
#define N 10
int main()
{
 float a[N]={0.974742, 0.0982212, 0.578671, 0.717988, 0.881543,
  0.0771773, 0.910513,0.576627, 0.506879, 0.629856};
 float sum=0, average, var=0, sd;
 int i;
 for (i=0;i<N;i++) sum=sum+a[i];
 average=sum/N;
 for (i=0;i<N;i++) var=var+pow( a[i]-average, 2);
 sd=sqrt((var)/(N-1.0));
 printf("Average= %f S.D.=%f\n", average, sd);
 return 0;
}
```

The output is

```
$ gcc prog.c -lm
$ ./a.out
Average= 0.595222 S.D.=0.310107
```

2. Regression analysis (curve fitting)

   As another example of arrays, regression analysis that can obtain the best fit curve for given experimental data is introduced. Table 2.6 shows N measured points.

Table 2.6: N measured points

X	x_1	x_2	x_3	...	x_N
Y	y_1	y_2	y_3	...	y_N

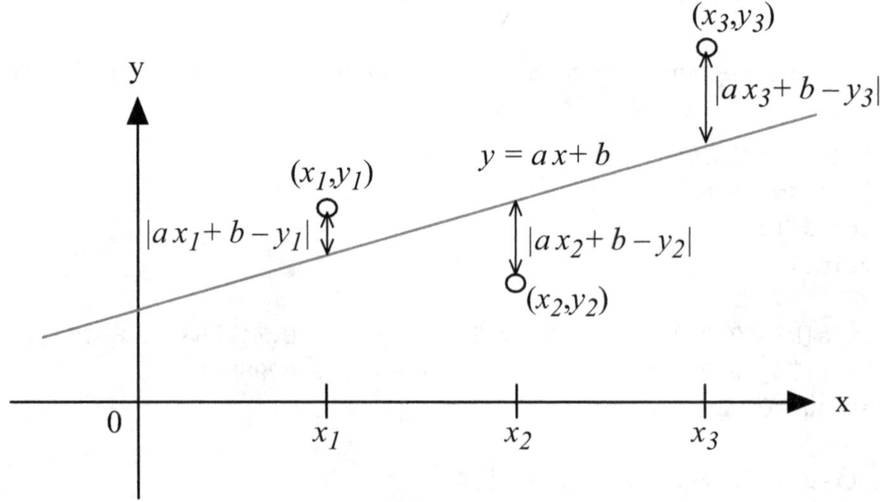

Figure 2.2: General regression analysis.

The best fit line that represents the above data points in the format of

$$y = ax + b$$

is sought. The two parameters, a (slope) and b (intercept with the y axis), need to be chosen to minimize an error. The error is defined as the difference between the measured value and predicted value. Since this value can be either positive or negative, it is necessary to make it positive-definite by squaring the nominal value. Therefore, the total error,[21] E^2, is the sum of the square of the difference at each point defined as

$$\begin{aligned} E^2 &\equiv (ax_1 + b - y_1)^2 + (ax_2 + b - y_2)^2 + \ldots + (ax_N + b - y_N)^2 \\ &= \sum_{i=1}^{N} (ax_i + b - y_i)^2. \end{aligned} \tag{2.8}$$

Choose a and b so that E^2 be minimized. This can be achieved by partially differentiating E^2 in Eq. (2.8) with respect to both a and b and set them to be 0 as[22]

$$\frac{\partial E^2}{\partial a} = 0, \quad \frac{\partial E^2}{\partial b} = 0.$$

This yields a set of two simultaneous equations for a and b as

$$\begin{aligned} 2\sum_{i=1}^{N} (ax_i + b - y_i)x_i &= 0, \\ 2\sum_{i=1}^{N} (ax_i + b - y_i)(+1) &= 0, \end{aligned}$$

or

$$\begin{aligned} \left(\sum_{i=1}^{N} x_i^2\right) a + \left(\sum_{i=1}^{N} x_i\right) b &= \sum_{i=1}^{N} x_i y_i, \\ \left(\sum_{i=1}^{N} x_i\right) a + \left(\sum_{i=1}^{N} 1\right) b &= \sum_{i=1}^{N} y_i, \end{aligned}$$

which can be solved using Cramer's rule as

$$\begin{aligned} a &= \frac{\begin{vmatrix} \sum_{i=1}^{N} x_i y_i & \sum_{i=1}^{N} x_i \\ \sum_{i=1}^{N} y_i & N \end{vmatrix}}{\begin{vmatrix} \sum_{i=1}^{N} x_i^2 & \sum_{i=1}^{N} x_i \\ \sum_{i=1}^{N} x_i & N \end{vmatrix}}, \\ b &= \frac{\begin{vmatrix} \sum_{i=1}^{N} x_i^2 & \sum_{i=1}^{N} x_i y_i \\ \sum_{i=1}^{N} x_i & \sum_{i=1}^{N} y_i \end{vmatrix}}{\begin{vmatrix} \sum_{i=1}^{N} x_i^2 & \sum_{i=1}^{N} x_i \\ \sum_{i=1}^{N} x_i & N \end{vmatrix}}. \end{aligned}$$

The following program implements this result using 10 data points:

[21] The dimension of E^2 is the same as a^2 and b^2. Hence, it is denoted as E^2.
[22] This method is called the least square method.

```
#include <stdio.h>
#define N 10
int main(){
 float x[N]={1,2,3,4,5,6,7,8,9,10},
  y[N]={-3, 2.9, 0.5, 3.0, 2.6, 5.2, 4.9, 4.5, 6.1, 7.2};
 float xysum=0.0, xsum=0.0, ysum=0.0, x2sum=0.0;
 float a, b, det;
 int i;
 for (i=0; i< N; i++)
 {
  xsum = xsum + x[i];
  ysum = ysum + y[i];
  xysum = xysum + x[i]*y[i];
  x2sum = x2sum + x[i]*x[i];
 }
 det=x2sum*N - xsum*xsum;
 a = (xysum*N-xsum*ysum)/det;
 b = (x2sum*ysum-xysum*xsum)/det;
 printf("The regression line is %f x + %f.\n", a, b);
 return 0;
}
```

The output is

```
$ gcc prog.c
$ ./a.out
The regression line is 0.863636 x + -1.359998.
```

It is also possible to apply the regression analysis to a curved line such as $y = ax^2 + bx + c$.

2.6.4 EXERCISES

1. Birthday paradox

   How likely is it that two persons in a group of N people have the same birthday?[23] Use the Monte Carlo method to estimate the probability.

[23]This problem is known as the birthday paradox and can be solved exactly by the probability theory. Mathematically, this probability is expressed as $p(N) = 1 - \frac{365!}{365^N (365-N)!}$. When $N = 23$, the probability is over 50%.

Suggested Approach:

(a) Prepare an integer array, `a[N]`, that holds birthday dates for `N` people.

(b) Generate a random number between 1 and 365 and assign the number to each element of `a[]`.[24]

(c) Compare `a[0]` with the rest of birthdays and if there is a match, get out of the loop, increment the counter by 1 and go to (b) (the next round of simulation).

(d) Compare `a[1]` with the rest of birthdays and if there is a match, get out of the loop, increment the counter by 1 and go to (b) (the next round of simulation).

(e) ....

(f) After n simulations, compute the value of **counter/n**.

To exit from the loop, use the `goto` statement as

```
 for (i=0;i<N-1;i++) for (j=i+1;j<N;j++)
   if(a[i]==a[j])
          {count++;goto ExitLoop;}

  ExitLoop:;
.....
```

Try n = 100, 1,000, 100,000.

2. Throw a die 10, 1,000 and 10,000 times and compute the average and the standard deviation.

   Template:

```
#include <stdio.h>
#include <stdlib.h>
#include <time.h>
#include <math.h>
int main()
{
  int i,a[10000];
  float sum=0, avg, std;
  srand(time(NULL));
```

[24]Use the % operator. Example: 300%7 returns the remainder of 300 divided by 7, i.e., 6.

```
    for (i=0;i<10000;i++) a[i]=....;

    for (i=0;i<10000;i++) sum+= ....;
    ...................;
    return 0;
}
```

2.7 FILE HANDLING

In this section, how C programs can interact with external files is explained.

2.7.1 I/O REDIRECTION (STANDARD INPUT/OUTPUT REDIRECTION)

UNIX shells (csh/tcsh/bash on most of UNIX platforms) and the DOS window have the capability of I/O redirection.[25] Instead of entering data from the keyboard and displaying the output on the computer screen, it is possible to re-direct input/output to/from other devices such as files, printers, etc. Table 2.7 summarizes the available options.

Table 2.7: I/O redirections

Notion	Meaning
$ program > filename	Output to file
$ program >> filename	Append output to file
$ program < filename	Get input from file

If `a.out` alone is entered from the keyboard, all the output from `a.out` is displayed on the screen. However, with `a.out > result.dat`, the output is saved to an external file, `result.dat`, and nothing is shown on the screen. You can examine the contents of `result.dat` by the `more` command.

```
$ gcc prog.c
$ ./a.out > result.dat
$ more result.dat
```

If the program requires that the input must come from an existing external file and the output must be saved to another external file, you can have both directions on the same line as

[25]This is operating system dependent and not a property of the C language.

```
$ gcc prog.c
$ ./a.out < data.dat > result.dat
$ more result.dat
```

The file, `data.dat`, contains the necessary data to be input to the program.

The following command (in a DOS window) creates a file, `filelist.dat`, that lists all the files in the current directory:[26]

```
c:\dir > filelist.dat
```

2.7.2 FILE HANDLING (FROM WITHIN A PROGRAM)

I/O redirection is operating system dependent (UNIX and DOS) and only available when the C program is run from a command line. It is not possible to use I/O redirection when the C program is run under GUI.

To write/read an external file from within a C program, the file must be opened first and must be closed after necessary operations on the file are done.

For opening/closing a file, the functions, `fopen()` and `fclose()`, with a special keyword `FILE`[27] (note the upper case) must be used. Use the following syntax. The file variable, `fp`, is a pointer (the topic in Section 2.8).

```
#include <stdio.h>
int main()
{
 FILE *fp;
 fp = fopen("filename","w");
 /*
   Write something to fp.
 */
 fclose(fp);
 return 0;
}
```

The function, `fopen()`, takes `a` (append), `w` (write), or `r` (read) as a possible argument. The following program opens an external file, `data.dat`, for writing and writes "`Hello, World!`" to the file.

[26]In the Windows system, a file, `prn`, cannot be used as the name for an external file. It is reserved for a printing device.
[27]This is a FILE pointer that keeps track of the memory location of the file.

```
#include <stdio.h>
int main()
{
 FILE *fp;
 fp=fopen("data.dat","w");
 fprintf(fp,"Hello, World!\n");
 fclose(fp);
 return 0;
}
```

The following program reads three floating numbers separated by a space from an external file, `data.dat`, and prints out their values on the screen:

```
#include <stdio.h>
int main()
{
 FILE *fp; float a,b,c;
 fp=fopen("data.dat","r");
 fscanf(fp,"%f %f %f", &a, &b, &c);
 printf("%f %f %f", a,b,c);
 fclose(fp);
 return 0;
}
```

Multiple files can be opened for writing/reading as

```
#include <stdio.h>
int main()
{
 FILE *fp1, *fp2;
 float a,b,c;
 fp1=fopen("data1.dat","w");
 fp2=fopen("data2.dat","w");
 fprintf(fp1,"This is the first file.\n");
 fprintf(fp2,"This is the second file.\n");
 fclose(fp1); fclose(fp2);
 return 0;
}
```

2.8 POINTERS

The concept of a pointer in C is probably the most challenging subject for a C learner. Pointers are available in C and C++ but not in any other programming languages. Your program looks just like what a C program should appear if pointers are used in many places.

However, there are only two new operators (& and *) that need to be learned in order to understand the pointer. The good news for scientific and engineering programming is that you can get around using a pointer for most of programming needs. The only instances that a pointer is needed for engineering/science are (1) working on matrices and vectors in linear algebra and (2) using a variable that is substituted for a function both of which will be explained in this section.

2.8.1 ADDRESS OPERATOR & AND DEREFERENCING OPERATOR *

Pointers are closely related to the hardware of a computer that runs a C program. When a C compiler reads a source code, the compiler maps each variable defined in the code to an appropriate location in RAM that holds the value of that variable. For example, consider the following program:

```
#include <stdio.h>
int main()
{
 float a =20.0, b=50.0;
 float *pa, *pb;
 pa=&a; pb=&b;
 return 0;
}
```

When the compiler reads the program above, it will make a table of memory mapping which may look like Table 2.8. Note that this is a fictitious machine for illustrative purposes.

In this fictitious machine, the variable `a` is mapped to the (absolute) memory location at `102` in RAM which holds the value of 20.0. Similarly, the variable `b` is mapped to the memory location at `150` that holds the value of 50.0.

You can find out the memory location (address) that holds the value of a given variable by placing an `&` (ampersand, called an address operator or a referencing operator) in front of the variable name. Thus in the example above, `a` represents 20.0 and `&a` represents 102. Run the following program:

```
#include <stdio.h>
int main()
```

Table 2.8: Memory map on a fictitious machine

Variable Name	Absolute Memory Address	Content
...	100	...
...	101	...
a	102	20.0
...	...	...
b	105	50.0
...	151	...
...	...	...
pa	200	102
...	...	...
pb	220	150

```
{
  int a=10;
  printf("Address of a=%d.\n", &a);
  return 0;
}
```

The compiler outputs a warning and the output from `a.out` is negative, which is wrong.

```
$ gcc prog.c
prog.c: In function 'main':
prog.c:5:10: warning: format '%d' expects argument of type 'int',
                       but argument 2 has type 'int *' [-Wformat=]
   printf("Address of a=%d.\n", &a);
          ^
$ ./a.out
Address of a=-1074704200.
```

This is because `&a` holds the address of a memory location which is larger than the maximum integer that can be handled by the compiler (=2147483647). The `%p` format instead of the `%d` format should be used to display a memory address in hexadecimal mode[28] correctly. Change `%d` to `%p` in `printf()` in the previous code. The output will be something like (each machine is different)

[28]In hexadecimal format, 16 letters (0–9 and a–f) are used to represent a number.

```
$ gcc prog.c
$ ./a.out
Address of a=0xbfa1e9a8.
```

Note that a number beginning with `0x` or `oX` is interpreted as a number in hexadecimal format. Compare the following two programs side by side using an array `a[3]`:

```
#include <stdio.h>
int main()
{
  int a[]={100,2,-56};
  printf("%p\n", &a[0]);
  printf("%p\n", &a[1]);
  printf("%p\n", &a[2]);
  return 0;
}
```

```
#include <stdio.h>
int main()
{
  double a[]={100,2,-56};
  printf("%p\n", &a[0]);
  printf("%p\n", &a[1]);
  printf("%p\n", &a[2]);
  return 0;
}
```

The output from the two programs may look like (each machine is different)

```
$ gcc prog1.c
$ ./a.out
0xbfc76660
0xbfc76664
0xbfc76668
```

```
$ gcc prog2.c
$ ./a.out
0xbf988c50
0xbf988c58
0xbf988c60
```

In the first (left) program, the array, `a[]`, is declared as an integer array while in the second (right) program, `a[]` is declared as a double precision array.

In the output from the first (left) program, the address of adjacent components is separated by 4 bytes which implies that the C compiler stores each integer number (`int`) using 4 bytes. On the other hand, in the output from the second (right) program, the address is incremented by 8 bytes which implies that the C compiler stores each double precision (`double`) number using 8 bytes.

2.8.2 PROPERTIES OF POINTERS

A pointer is a variable just like `a` or `b` above. The only difference is that it stores the "address of another variable." Therefore, if pa is a pointer, the content of pa is not a regular number such as 20.0 or 50.0 but more like a large number such as `0xbf988c50` as in the examples above.

You must declare a pointer variable the same way you declare any regular variable but with an asterisk (*) preceding the variable name as

```
#include <stdio.h>
int main()
{
  float a=20.0;
  float *pa;
  pa=&a;
  printf("%p\n", pa);
  printf("%p\n", &a);
  return 0;
}
```

The output (each machine is different) looks like

```
$ gcc prog.c
$ ./a.out
0xbfb0f214
0xbfb0f214
```

In the program above, `float *pa` declares that pa is a pointer pointing to a `float` variable. Note that a pointer itself is always a large integer (memory address location) so the type of a pointer (`float` in the example above) merely implies that the variable pointed by the pointer pa is of `float` type. The `pa=&a;` statement means that the address of a is assigned to pa.

Sometimes within the program you want to examine what value the pointer actually refers to, i.e., you want to know the content of the variable pointed by the pointer. Let's look at the following program:

```
#include <stdio.h>
int main()
{
  float a=20.0;
  float *pa;
  pa=&a;
  printf("%f\n", *pa);
  printf("%f\n", a);
  return 0;
}
```

The output is

```
$ gcc prog.c
$ ./a.out
20.000000
20.000000
```

In the program above, * (asterisk—known as the dereferencing operator) before a pointer variable works the same way as using a directly. Therefore, if pa is a pointer pointing to a floating variable a, the use of a and *pa within a program is identical. This way, you can effectively change the content of a without directly mentioning a.

Pointers can be incremented or decremented just as any other variables. The difference between the increment of pointers and the increment of regular variables is that the value of increment for the pointer depends on the type of the variable the pointer is pointing to. Take the following example:

```
#include <stdio.h>
int main()
{
  float  a[3]={1.0, 2.0, 3.0}, *pa=&a[0];
  double b[3]={1.2345670, 2.009876555, 3.14159265}, *pb=&b[0];
  printf("float  %15p%15p%15p\n", pa, pa+1, pa+2);
  printf("double %15p%15p%15p\n", pb, pb+1, pb+2);
  return 0;
}
```

The output looks like (each machine is different)

```
$ gcc prog.c
$ ./a.out
float        0xbf8135e4     0xbf8135e8     0xbf8135ec
double       0xbf8135f0     0xbf8135f8     0xbf813600
```

In the program above, as a double precision variable occupies more memory space (8 bytes) than a single precision variable (4 bytes), the increment of the double precision pointer, pb, advances the memory location by 8 bytes while the increment of the single precision pointer, pa, advances the memory location by 4 bytes even though the increment for both pointers is 1.

The C language uses pointers extensively for the following reasons:

- The only way to modify the content of arguments in a function call is to use pointers.
- Computer hardware can be directly controlled by pointers.
- Matrices and vectors in linear algebra are actually pointers.

2.8.3 FUNCTION ARGUMENTS AND POINTERS

One of the usages of pointers is to modify the content of a variable (parameter) passed through a function call. Suppose you want to write a function, `tentimes()`, that accepts a variable, `a`, as a parameter and multiply 10 by `a`. Such a program may look like

```
#include <stdio.h>
void tentimes(float a)
{
  a = 10.0*a;
}
int main()
{
  float b=20;
  tentimes(b);
  printf(" b = %f\n", b);
  return 0;
}
```

The output is

```
$ gcc prog.c
$ ./a.out
  b = 20.000000
```

Unfortunately, this program does not work as intended. The content of `b` was not modified even with the `a = 10.0*a` statement in the definition of `tentimes()`.

There is nothing wrong with this program as this is how a function call in C works. When a function is called with an argument (i.e., `tentimes(b)`), a copy of the value of the argument (i.e., `b`) is made, that value is passed to the function and the actual argument variable `b` is never altered. All that the function (i.e., `tentimes()`) can see from the function `main()` is the copied value of the parameter (i.e., 20) and this information does not have anything to do with the variable itself (i.e., `b`). This effectively makes it impossible to modify the value of arguments passed to functions. This way of passing a parameter in C functions is called the *call by value* method.

A fix is to use pointers. Instead of passing a copy of the value of the variable to the function, if the address of that variable is passed to the function, the function can find where the value of the variable is stored, go to that address, and modify the content at that address any way it wants. This is called the *call by reference* method. The program above can be now modified as

```
#include <stdio.h>
void tentimes(float *a)
{
 *a = 10.0**a;
}
int main()
{
 float b=20;
 tentimes(&b);
 printf(" b = %f\n", b);
 return 0;
}
```

The output from this program is

```
$ gcc prog.c
$ ./a.out
 b = 200.000000
```

This output is what was expected. Note that the dummy variable, `a`, is declared as a pointer (`*a`) and instead of `b`, the address of `b` (`&b`) is passed to the function, `tentimes()`. In the function, `tentimes()`, `*a` represents the content of the variable pointed by `a`.[29]

As an example of this concept, the following program uses a function, `swap()`, that takes two pointers as arguments and exchanges the values of the two variables pointed by the pointers:

```
#include <stdio.h>
void swap(float *pa, float *pb)
{
 float tmp;
 tmp=*pa;
 *pa=*pb;
 *pb=tmp;
}
int main()
{
 float a=10.0, b=50.0;
 printf("Old a = %f and old b = %f\n",a,b);
 swap(&a,&b);
```

[29] `10.0**a` is the product between 10.0 and the value pointed by `a`, i.e., `*a`.

```
  printf("New a = %f and new b = %f\n", a,b);
  return 0;
}
```

The output is

```
$ gcc prog.c
$ ./a.out
Old a = 10.000000 and old b = 50.000000
New a = 50.000000 and new b = 10.000000
```

2.8.4 POINTERS AND ARRAYS

Handling matrices and vectors in linear algebra is the most relevant use of pointers for scientific and engineering programming. When an array, a[3], is declared as

```
#include <stdio.h>
int main()
{
float a[3]={1.0, 2.0, 3.0};
return 0;
}
```

the C compiler actually interprets a as a pointer and reserves appropriate memory space. In the program above, an array a is a pointer pointing to the 0-th element of the array, a[0], while a[0], a[1], and a[2] behave just like regular variables. The content of a is the address which points to a[0]. Since a is a pointer, dereferencing the array name (*a) will give the 0-th element of the array, i.e., a[0]. This gives us a range of equivalent notations for accessing arrays.

In Table 2.9, *(a+2) means that the value (address) of the pointer, a, is advanced by two units and the content at that address is referenced which is equivalent to the value of a[2].

Table 2.9: Accessing array elements in two different ways

Array Access	Pointer Equivalent
a[0]	*a
a[1]	*(a+1)
a[2]	*(a+2)

Since an array is a pointer, it is possible to pass an array to a function, and modify any element of that array. Here is an example:

```
#include <stdio.h>
void twice(float *a)
{
 int i;
 for (i=0; i<3; i++) a[i]=2*a[i];
}
int main()
{
 float b[3]={1.0, 2.0, 3.0};
 int i;
 twice(b);
 for (i=0; i<3; i++) printf("%f\n", b[i]);
 return 0;
}
```

The output from the program is

```
$ gcc prog.c
$ ./a.out
2.000000
4.000000
6.000000
```

The program above is to use a function that takes an array as an input and doubles all the elements in that array. Previously, it was noted that the only way to modify the argument in a function is to use a pointer. This principle also works for arrays as the name of an array is a pointer. Therefore, when passing an array to a function, the address operator, &, before the array name is not needed. The following program does the same as the one above:

```
#include <stdio.h>
void twice(float a[3])
{
 int i;
 for (i=0; i<3; i++) a[i]=2*a[i];
}
int main()
```

```
{
 float b[3]={1.0, 2.0, 3.0};
 int i;
 twice(b);
 for (i=0; i<3; i++) printf("%f\n", b[i]);
 return 0;
}
```

2.8.5 FUNCTION POINTERS

There is another type of pointer that can point to a function instead of a variable. Such a pointer is called a *function pointer*. Using a function pointer, it is possible to use a variable representing different functions. A problem may arise when it is necessary to perform some operation (such as numerical integration) for many different functions. Without using a pointer to a function, such a program needs to be written to repeat the operations for as many times as the number of different functions. With the use of a function pointer, the program can be written such that the name of a function can be handled just like a variable name in which the variable for the function name can be substituted for any name of the actual function.

When a function is declared such as `float myfunc()`, the name of the function, `myfunc`, itself is actually a pointer pointing to a memory location where the routine of the function code begins.[30]

A function pointer can be declared as

```
type_of_function (*func_name)(type_of_argument)
```

where `type_of_function` is the type of a function to which the function points, `func_name` is the name of the function pointer and `type_of_argument` is the type of the argument in `func_name`. For example, a function pointer, `float (*foo)(float, float)`, can be pointed to an actual function, `float f1(float x, float y)`.

In the following code, `func()` is a function pointer which points to an actual `double` type function that has a `double` type variable as the argument.

```
#include <stdio.h>
#include <math.h>
double f1(double x)
 {return x ; }
double f2(double x)
```

[30]This is similar to an array. The array name itself is a pointer pointing to the address of the first element of the array.

```
 {return x*x ; }

int main()
{
 double (*func)(double);

 func=&f1;
  printf("%lf\n", func(2)) ;
 func=&f2;
  printf("%lf\n", func(2));
 func=&cos;
  printf("%lf\n", func(3.141));
 return 0;
}
```

The output is

```
$ gcc prog.c -lm
$ ./a.out
2.000000
4.000000
-1.000000
```

As `func()` is a function pointer, a statement such as `func=&cos` assigns the address of the cosine function defined in `<math.h>` to `func`. After this statement, `func()` and `cos()` are identical.

2.8.6 SUMMARY

- The asterisk (*) in C has the following meanings:

  1. Multiplication if used as a binary operator (`a*b`).
  2. The value at the address contained in a pointer if used as a unary operator (`*pa`) (dereferencing operator).
  3. Declaration of a variable as a pointer used in the declaration statement (`int *pa`).

- The ampersand symbol (&) in C has the following meanings:

  1. The address of a variable if used as a unary operator (`&a`) (referencing operator).
  2. Logical OR if used as a binary operator (`a==1.0 && b==2.0`).
  3. Bitwise OR between two binary numbers (`a & b`).

- Pointers must be used if the values of arguments in a function need to be changed. This is the reason why `scanf()` requires & but not `printf()`. The value of the variable to be passed to `scanf()` is entered from the keyboard. Hence, it is necessary to pass the address of that variable to `scanf()` so that the content of the variable can be modified.

2.8.7 EXERCISES

1. We want to write a function, `circle`, which takes the radius of a circle from the main function and assign the area of the circle to the variable, `area`, and the perimeter of the circle to the variable, `perimeter`. Since the contents of the variables are to be modified by a function, it is necessary to use pointers. Use the following template and complete your program.

```
#include <stdio.h>
void circle(float r, (fill in your code))
{
 (fill in your code...);
}
int main()
{
 float r, area, perimeter;
 printf("Enter radius = "); scanf("%f", &r);
 circle(r, &area, &perimeter);
 printf("r=%f area=%f peri=%f\n", r, area, perimeter);
 return 0;
}
```

2. Write a function that takes two floating variables, a and b, and rearrange them in ascending order, i.e.,

```
#include <stdio.h>
void reorder(float *pa, float *pb)
{
 (your code here)
}
int main()
{
 float a=15, b=-6;
 reorder(&a, &b);
```

```
 printf("%f %f\n", a, b);
 return 0;
}
```

shows

```
-6 15
```

3. Remember that an array is also a pointer as exemplified by

```
#include <stdio.h>
int main()
{
 int a[5]={1,2,3,4,5};
 int i;
 for (i=0;i < 5;i++) printf("%d\n", a[i]);
 for (i=0;i < 5;i++) printf("%d\n", *(a+i));
 return 0;
}
```

so `a[i]` and `*(a+i)` are identical. Using this idea, write a program to compute the average of all the elements for the following array:

```
a[20]={0.228952, 0.568418, 0.820277, 0.117099, 0.755212,
0.509299, 0.572073, 0.224526, 0.852861, 0.0612133, 0.175636,
0.568243, 0.0100543, 0.702012, 0.0345108, 0.146549, 0.189951,
0.144139, 0.261263, 0.474034};
```

2.9 STRING MANIPULATION

2.9.1 HOW TO HANDLE A STRING OF CHARACTERS (TEXT)

A text (a string of characters) in C is handled as a set of individual characters, i.e., an array of characters. Therefore, an array has to be used to represent a text (string). Since any array variable is a pointer as discussed in Section 2.8, the variable that represents a string is a pointer as well.

Compare the following programs side by side:

```
#include <stdio.h>
int main()
{
 float a=2.0;
 printf("%f\n",a);
 return 0;
}
```

```
#include <stdio.h>
int main()
{
 char b='A';
 printf("%c\n",b);
 return 0;
}
```

In the left program above, the `float` variable, a, represents a single value, 2.0. In the right program above, the char variable, c, represents a single character, A.

```
#include <stdio.h>
int main()
{
 float a[]={2.0, 3.0, 4.0, 5.0};
 printf("%f\n",a[0]);
 return 0;
}
```

```
#include <stdio.h>
int main()
{
 char b[]="Hello!";
 printf("%c\n",b[0]);
 return 0;
}
```

In the left programs above, an array, a, represents a set of four numbers, `2.0, 3.0, 4.0, 5.0` and the first number is printed. In the right programs above, an array, b, represents a string of six characters, "`Hello!`," and the first character is printed.

```
#include <stdio.h>
int main()
{
 float *a, b[]={1.0,2.0};
 a=b;
 printf("%f\n",a[0]);
 return 0;
}
```

```
#include <stdio.h>
int main()
{
 char *a, b[]={'H','E','L',
               'L','O','!','\0'};
 a=b;
 printf("%c\n",a[0]);
 return 0;
}
```

In the left program above, a pointer variable, a, points to the address of `a[0]`. In the right program, a pointer variable, a, points to the address of the first character, "H." Instead of using `b[]={'H', 'E', 'L', 'L', 'O', '!', '\0'}`, a direct assignment of `b[]="HELLO!"` can be used with the double quotation mark ("). Note that a special character `'\0'` needs to be placed to mark the end of the string when we initialize each character individually.

Use the %s format to represent the entire string instead of the %c format that represents only one character.

```
#include <stdio.h>
int main()
{
 char *a;
 a="Hello, World!";
 printf("%s\n",a);
 return 0;
}
```

The output is

```
$ gcc prog.c
$ ./a.out
Hello, World!
```

As is seen in the examples above, a single quotation mark (`'`) and a double quotation mark (`"`) work differently. The single quotation mark must be used for one character while the double quotation mark must be used for a string of characters. For example, `"ABC"` (double quotation) is for a string of characters and is therefore an array (pointer). On the other hand, `'A'` (single quotation) is for a single character.

Consider the following program:

```
#include <stdio.h>
int main()
{
 char s[4] = "ABC";
 printf("%s\n", s);
 return 0;
}
```

You may wonder why the element number of the array, `s`, is 4, not 3. The C compiler automatically adds a special character NULL (ASCII code 0, commonly denoted as "\0") to the end of the string to indicate that that is the end of the character array so that the element number of a character array is always the number of characters plus 1.

To read a string from the standard input (i.e., keyboard), use the following example:

```
#include <stdio.h>
int main()
{
```

```
  char str[100];
  printf("Enter a word = ");
  scanf("%s", str);
  printf("%s\n",str);
  return 0;
}
```

The output may look like

```
$ gcc prog.c
$ ./a.out
Enter a word = Good morning
Good
```

Note that there is no "&" before `str` in the `scanf()` function as `str` is already a pointer. Also, only "`Good`" is printed even though "`Good morning`" was entered. This is because the format, `%s`, in `scanf()`, reads input until a space is entered (a word). If two strings (two words) are to be read, `"%s %s"` must be used. The C compiler automatically adds "`\0`" after the end of the last character in the string.

2.9.2 STRING COPY/COMPARE/LENGTH

To copy a string to another string or to compare a string against another string, it is best to use functions, `strcpy()` and `strcmp()`, available in `<string.h>`.

```
#include <stdio.h>
#include <string.h>
int main()
{
 char c1[] = "ABCDE", c2[6];
 strcpy(c2,c1);
 printf("%s\n",c2);
 return 0;
}
```

The output is

```
$ gcc prog.c
$ ./a.out
ABCDE
```

In the program above, the function, `strcpy(c2,c1)`, copies the pointer, `c1`, to another pointer, `c2`. Note `c2[6]` instead of `c2[5]` as the NULL character must be added after the end of the string.

To compare two strings, use `strcmp()` available in `<string.h>`:

```
#include <stdio.h>
#include <string.h>
int main()
{
 char s[100];
 printf("Enter \"ABCDEF\"");
 scanf("%s", s);
 if (strcmp(s,"ABCDEF") == 0) printf("ABCDEF was entered correctly.\n");
 else printf("Wrong. %s was entered.\n",s);
 return 0;
}
```

In the program above, the `strcmp()` function takes two strings as the arguments and returns 0 if the two strings match. Note that you can output " (double quotation mark) by "escaping" it using the backslash character.

To find the length of the string, use the `strlen()` function:

```
#include <stdio.h>
#include <string.h>
int main()
{
 char c[50];
 printf("Enter string = ");
 scanf("%s", c);
 printf("You entered %s\n", c);
 printf("Its length is %d\n", strlen(c));
 return 0;
}
```

The output is

```
$ gcc prog.c
$ ./a.out
Enter string = Good afternoon.
```

```
You entered Good
Its length is 4
```

Again, only "Good" was read as the format, %s, reads a string until a space is entered (a word).

2.10 COMMAND LINE ARGUMENTS

2.10.1 ENTERING COMMAND LINE ARGUMENTS

The traditional way of executing a C program after successful compilation is to issue the program name at the system prompt as in the following example:

```
$ gcc prog.c -lm -o prog
$ ./prog
(interactive session)
```

Instead of entering necessary information during the interactive session (using `scanf()` and `printf()` functions), you can enter necessary parameters at the same time you enter the program name as

```
$ gcc prog.c -lm -o prog
$ ./prog 3421 8756
(executes program to manipulate 3421 and 8756 and prints results)
```

This is called "command line arguments" (also called "command line parameters") and provides a handy way to pass necessary arguments to the `main()` function without user interaction.

Recall that the function, `main()`, is the function in C that is to be executed first. Other than that, it is an ordinary function in C just like any other functions so that it must have an argument part. The function, `main()`, actually takes two arguments. The first argument is the number of command line arguments including the program name itself and the second argument (an array of strings) stores each command line argument entered after the program name. The syntax of the arguments in `main()` is

```
int main(int argc, char *argv[])
```

The first argument, `argc`, is of integer type and is assigned the number of command line arguments including the command name itself. The second argument, `argv`, is a pointer to an array of strings that stores each of the command line arguments. Note that `argv[]` is an array so that it can take multiple command line arguments.[31]

[31]Actually, `argv[]` is a pointer to another pointer (an array of another array) as each element of `argv[]` is a string of characters.

Consider the following program:

```
#include <stdio.h>
int main(int argc, char *argv[])
{
 int i;
 printf("Number of arguments = %d\n", argc);
 for(i=0; i<argc; i++)  printf("%d: %s\n", i, argv[i]);
 return 0;
}
```

Execute the program above with command line arguments as in the following example:

```
$ gcc prog.c
$ ./a.out I hate C language.
Number of arguments = 5
0: ./a.out
1: I
2: hate
3: C
4: language.
```

The string "`./a.out`" is stored in `argv[0]`, the string "`I`" is stored in `argv[1]`, etc.

Note that each of the command line arguments is entered as a string. For example, when a number 4 is entered, it is stored as a character "4" (ASCII code 52), not as the numeric value of 4. If you want the program to interpret the entered parameters as numeric values, not a string of characters, use `atoi()` (ASCII to INTEGER) or `atof()` (ASCII to FLOAT) available in `<stdlib.h>`.

```
#include <stdio.h>
#include <stdlib.h>
int main(int argc, char *argv[])
{
 printf("%d\n", atoi(argv[1]));
 return 0;
}
```

The output looks like

```
$ gcc prog.c
$ ./a.out 2018
2018
```

The following program computes the sum of all the numbers entered as command line arguments:

```
#include <stdio.h>
#include <stdlib.h>
int main(int argc, char *argv[])
{
 float sum=0.0; int i;
 for (i=1;i<argc;i++)  sum=sum+atof(argv[i]);
 printf("The sum is %f.\n" , sum);
return 0;
}
```

The output looks like

```
$ gcc prog.c
$ ./a.out 1 2 3 4 5 6 7 8 9 10
The sum is 55.000000.
```

2.10.2 EXERCISES

1. Write a program using the concept of command line arguments to solve a quadratic equation, $ax^2 + bx + c = 0$, by entering a, b, and c as command line arguments, i.e.,

   ```
   $ ./a.out 3 -1 -1
   ```

   will return

   ```
   x1= -0.434259, x2 = 0.767592.
   ```

   by solving $3x^2 - x - 1 = 0$.

2.11 STRUCTURES

2.11.1 MIXTURE OF DIFFERENT TYPES OF VARIABLES

An array is a single variable that can represent many elements. However, all the elements in an array must be of the same type. What if you want to have a single variable that represents different types of elements such as a mixture of integers, floating numbers and strings?

A *structure* is a collection of one or more variables under a single name. These variables can be of different types, and are selected from the structure by name. A structure is a convenient way of grouping several pieces of related information together.

For example, a structure called `student` can be created which represents the school records of students such as the ID, the midterm score, the final score and the final grade. The following program exemplifies the concept of structures:

```
#include <stdio.h>
 struct student{
       char *ID;
       int Midterm;
       int Final;
       char Grade;
};
int main()
{
 struct student smith={"1000123456", 89, 98, 'A'},
                doe={"1000123457", 45, 53, 'F'};
 printf("%s\n", smith.ID);
 printf("%d\n", doe.Midterm);
 doe.Grade='D';
 printf("%c\n", doe.Grade);
 return 0;
}
```

The output is

```
$ gcc prog.c
$ ./a.out
1000123456
45
D
```

In the program above, a structure called student is defined having four members (ID, Midterm, Final, Grade) which is a mixture of different types. The first member, ID is a string (hence, a pointer), the second and third members, Midterm and Final, are integers and the last member, Grade, is a character. Two variables, smith and doe, are declared as of student type and initialized. A member in a structure can be accessed by using the dot (.).

You can also define an array of structures as

```
#include <stdio.h>
struct student{
        char *ID;
        int Midterm;
        int Final;
        char Grade;
};
int main()
{
 struct student myclass[15];
 int i;
 myclass[0].ID="10000123212";
 myclass[0].Grade='C';
 (.....)
 myclass[14].Grade='B';

 (.....)
 return 0;
}
```

It is also possible to use a pointer to a structure as

```
#include <stdio.h>
struct student{
        char *ID;
        int Midterm;
        int Final;
        char Grade;
};
int main()
{
  struct student Smith={"David Smith", 12, 45, 'F'}, *ptr;
```

```
  ptr = &Smith;
/*
..................
*/

  printf("%s %d %d %c\n", ptr->ID,     ptr->Midterm,
                          ptr->Final, ptr->Grade);
  return 0;
}
```

Note that a member in a structure can be referred by using ->.

Finally, by using `typedef`, you can define a structure and declare a variable as that `structure` type just like `integer` or `float` without preceding `struct`.

```
#include <stdio.h>
typedef struct {
        char *ID;
        int Midterm;
        int Final;
        char Grade;
       } student;

int main()
{
  student Jones={"Jones", 12, 45, 'F'}, *ptr;
  ptr = &Jones;
/*
..................
*/

 printf("%s\n", ptr->ID);
 return 0;
}
```

The concept of `structure` is extended to the concept of `class` which plays an essential role in C++ (and Java).

One of useful applications of structure in scientific/engineering computation is complex numbers. The C language does not support complex numbers[32] but it is easy to implement

[32]C++ has a complex number class.

complex numbers using structures. The following program defines a complex type $(a + bi)$ and computes addition of two complex numbers:

```
#include <stdio.h>
typedef struct
 {float Real; float Im;} Complex;
Complex ComplexAdd(Complex z1, Complex z2)
{
 Complex z;
 z.Real = z1.Real + z2.Real;
 z.Im = z1.Im + z2.Im;
 return z;
}
int main()
{
 Complex z1, z2, z;
 printf("Enter real and imaginary parts of z1 separated by space = ");
 scanf("%f %f", &z1.Real, &z1.Im);
 printf("Enter real and imaginary parts of z2 separated by space = ");
 scanf("%f %f", &z2.Real, &z2.Im);
 z = ComplexAdd(z1, z2);
 printf("%f + %f I \n", z.Real, z.Im);
 return 0;
}
```

The output is

```
$ gcc prog.c
$ ./a.out
Enter real and imaginary parts of z1 separated by space = 2 3
Enter real and imaginary parts of z2 separated by space = -1 4
1.000000 + 7.000000 I
```

Using `typedef`, the structure, `Complex`, behaves just like `int` or `float` and variables can be declared using `Complex` as `Complex z1, z2;`.

2.11.2 EXERCISES

1. Following the last example above, write a program to do complex division between two complex numbers, i.e., z_1/z_2. Using the program, compute z_1/z_2 where $z_1 = 2.12 + 1.21i$ and $z_2 = -2.8 + 7.8i$.

As a template, the following program computes the product of two complex numbers:

```
#include <stdio.h>
typedef struct
 {float Real; float Im;} Complex;
Complex ComplexMultiply(Complex z1, Complex z2)
{
 Complex z;
 z.Real = z1.Real*z2.Real - z1.Im*z2.Im;
 z.Im = z1.Real*z2.Im + z1.Im*z2.Real;
 return z;
}
int main()
{
 Complex z1, z2, z;
 z1.Real=0.25; z1.Im=-3.1412;
 z2.Real=0.98; z2.Im=1.655;
 z=ComplexMultiply(z1,z2);
 printf("The product of z1 * z2 = %f  + %f I.\n",
    z.Real, z.Im);
 return 0;
}
```

The output is

```
$ gcc prog.c
$ ./a.out
The product of z1 * z2 = 5.443686  + -2.664626 I.
```

PART II

Numerical Analysis

Now that the basic syntax of the C language has been explained, you are ready to write C programs to solve many problems arising in engineering and science. In Part II, numerical methods for solving nonlinear equations, a set of simultaneous equations and ordinary differential equations as well as numerically differentiating and integrating given functions are discussed. Solving these equations analytically requires advanced levels of mathematics. However, solving equations numerically often only involves intuitive or visual interpretation and not necessarily requires higher mathematics.

Although all the essential topics in numerical analysis are elucidated, it is not possible to cover every single aspect of numerical analysis. For a comprehensive reference, *Numerical Recipes in C*[33] is suggested.

[33]Press et al., *Numerical Recipes in C*, Cambridge, Cambridge University Press, 1996.

CHAPTER 3

Note on Numerical Errors

In Part I, the data type, `float`, was used for all real numbers where 4 bytes are allocated for each floating number. A `float` variable can represent 10^{-38} to 10^{38} which covers most of the practical range. However, this range is translated into 6 to 8 decimal digits of precision and many problems in science and engineering require more precision that this range. Consider the following examples:

1.

```
#include <stdio.h>
int main()
{
 float s=0.0;int i;
 for (i=0;i <10000;i++) s=s+0.1;
 printf("%f\n",s);
 return 0;
}
```

The intention of the program is to add 0.1 for 10,000 times. The result should be 1,000. However, the program outputs the following result:

```
$ gcc prog.c
$ ./a.out
999.902893
```

The output is not 1,000 but 999.902893 which is almost 1,000 but when precision is critical, this result is not acceptable. The error in this example comes from conversion between decimal numbers and binary numbers. The decimal numbers in the source code are converted to the binary numbers with a possibility of conversion error and the binary numbers are converted back to the decimal numbers adding another conversion error.[1]

[1]For example, conversion of decimal 0.1 to binary is a recurring binary **0.010011001100...** and conversion of binary **0.0001100110011001100** to decimal is **0.0999985**.

2.
```
#include <stdio.h>
int main()
{
 float a,b,c;
 a=123.45678;
 b=123.45655;
 printf("%f\n",a-b);
 return 0;
}
```

As this program subtracts 123.45655 from 123.45678, the result should be 0.00023. However, the output from the program is not what it is supposed to be.

```
$ gcc prog.c
$ ./a.out
0.000229
```

The output is not 0.00023 but 0.000229. The error may seem small but again if precision is required, this is not acceptable. The error in this code comes from subtracting a number from another number which is so close resulting in the loss of the significant figures (also called a cancellation error).

Both types of errors are inevitable and there is no way to completely eliminate such errors. However, possible harm can be minimized by using `double` instead of `float` for floating numbers. When a number is declared as `double`, it is assigned 8 bytes and effectively increases the valid range. While the range of a `float` variable is $\pm 10^{-38} \sim 10^{38}$ with seven significant digits, the range of a `double` variable is $\pm 10^{-308} \sim 19^{308}$ with 15 significant digits.

In the previous programs, change `float` to `double` and also change `%f` to `%lf` (long float) in the `printf()` function.

1.
```
#include <stdio.h>
int main()
{
 double s=0.0;int i;
 for (i=0;i<10000;i++) s=s+0.1;
 printf("%lf\n",s);
```

```
 return 0;
}
```

```
$ gcc prog.c
$ ./a.out
1000.000000
```

2.

```
#include <stdio.h>
int main()
{
 double a,b,c;
 a=123.45678;
 b=123.45655;
 printf("%lf\n",a-b);
 return 0;
}
```

```
$ gcc prog.c
$ ./a.out
0.000230
```

Another example of cancellation error is found in the innocent looking quadratic equation expressed as

$$ax^2 + bx + c = 0,$$

whose two roots are given as

$$x = \frac{-b \pm \sqrt{D}}{2a}, \quad D = \sqrt{b^2 - 4ac}. \tag{3.1}$$

Consider this seemingly trivial equation

$$x^2 + 200000x - 3 = 0,$$

where the exact solutions are

$$x_1 = -200000, \quad x_2 = 0.000015.$$

The coding for this equation is straightforward as

```
#include <stdio.h>
#include <math.h>
int main()
{
 float a, b, c, disc,x1, x2;
 a=1.0;b=200000;c=-3;
 disc=b*b-4*a*c;
 x1=(-b-sqrt(disc))/(2*a);
 x2=(-b+sqrt(disc))/(2*a);
 printf("x1 = %f, x2 = %f\n",x1,x2);
 return 0;
}
```

The output is

```
$ gcc prog.c -lm
$ ./a.out
x1 = -200000.000000, x2 = 0.000000
```

Apparently, x_2 is wrong although x_1 is correct. From Eq. (3.1), the discriminant is

$$\begin{aligned} D &= 20000^2 - 4 \times 1.0 \times (-3) \\ &= 40000000012, \\ \sqrt{D} &= 200000.00003. \end{aligned}$$

On the other hand, $b = 200000.0$. Therefore, a cancellation error occurs when b is subtracted from $\sqrt{D}$ as they are so close. Again, this can be avoided by using `double` as

```
#include <stdio.h>
#include <math.h>
int main()
{
 double a, b, c, disc,x1, x2;
 a=1.0;b=200000;c=-3;
 disc=b*b-4*a*c;
 x1=(-b-sqrt(disc))/(2*a);
 x2=(-b+sqrt(disc))/(2*a);
 printf("x1 = %lf, x2 = %lf\n",x1,x2);
 return 0;
}
```

The output is

```
$ gcc prog.c -lm
$ ./a.out
x1 = -200000.000015, x2 = 0.000015
```

In numerical analysis, do not use `float` but use `double` exclusively. As a variable declared as `double` occupies double the memory space (8 bytes) of a `float` variable (4 bytes), the size of the resulting executable increases but that's a small price to pay. If infinite precision is desirable, symbolic capable systems such as *Mathematica*[2] and *Maple* should be used.

[2]Wolfram, *The Mathematica Book*, Version 4, Cambridge University Press, 1999.

CHAPTER 4

Roots of $f(x) = 0$

In this chapter, numerical solutions for a single equation, $f(x) = 0$, are sought. The function, $f(x)$, can be a polynomial function or any nonlinear function of x.

The *fundamental theorem of algebra* states that an n-th order polynomial equation has n roots including complex roots. This does not mean that all the roots for polynomial equations can be obtained analytically in closed form. In fact, a fifth-order polynomial equation and beyond has no formula for their roots in closed form.[1]

Two important algorithms to numerically solve a single equation, $f(x) = 0$, are explained in this chapter. They are (1) the bisection method and (2) Newton's method. The bisection method is guaranteed to obtain at least one root while Newton's method provides a faster algorithm to obtain roots.

4.1 BISECTION METHOD

The bisection method is based on the *mean value theorem* which states that

If $f(x)$ is continuous over $x_1 \le x \le x_2$ and $f'(x)$ exists over $x_1 < x < x_2$ and $f(x_1)f(x_2) < 0$, there is at least one point x between x_1 and x_2 such that $f(x) = 0$.

As shown in Figure 4.1, if there is at least one zero for $f(x) = 0$ between x_1 and x_2, the product of $f(x_1)f(x_2)$ must be negative so that the curve crosses the x axis. Once such an interval, $[x_1, x_2]$, is identified, the next step is to divide this interval in half and repeat the test of whether $f(x_1)f(x_2) < 0$ is held with x_2 taken as $(x_1 + x_2)/2$ (middle point). If $f(x_1)f(x_2) < 0$ is held, then, the zero must be in the first interval. Otherwise, take the second interval as the interval in which the zero exists. Now that the interval in which there is at least one root is halved, repeat this procedure for a number of times until the width of the interval is sufficiently small or $f(x) = 0$ where x is the middle point chosen.

As shown in Figure 4.2, this procedure can be summarized as follows:

1. Choose x_1 and x_2 such that $f(x_1)f(x_2) < 0$.
2. Set $x_3 \leftarrow (x_1 + x_2)/2$.
3. If $f(x_1)f(x_3) < 0$, then, set $x_2 \leftarrow x_3$.
4. Else set $x_1 \leftarrow x_3$.

[1]This theorem known as the Galois theory was proven by a French mathematician, Évariste Galois (1811–1832), pronounced as *gal-wa*, when he was 19.

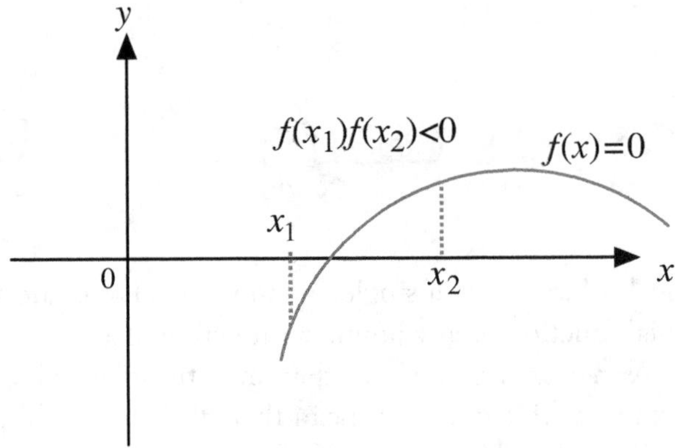

Figure 4.1: Bisection method.

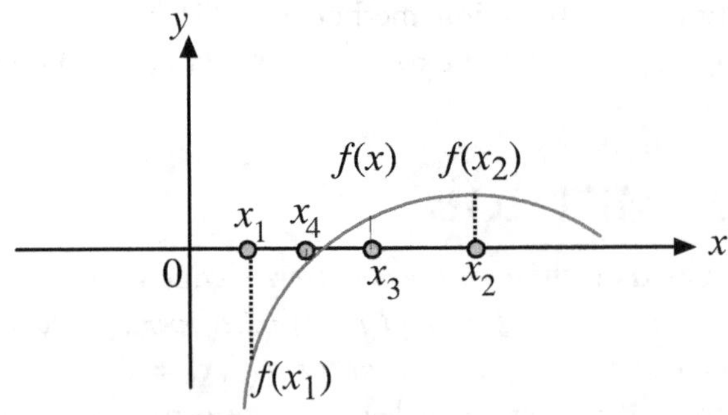

Figure 4.2: Algorithm of bisection method.

5. Until $|x_1 - x_2| < \epsilon$ (small threshold) or $f(x_3) = 0$, repeat 2–4.

The following C code is implementation of the bisection method for $x^2 - 2 = 0$. By solving this equation, an approximation to $\sqrt{2}$ can be obtained.

```
/* Compute the square root of 2 */
#include <stdio.h>
#include <math.h>
#define EPS 1.0e-10
#define N 100

double f(double x)
{
```

```
 return pow(x,2)-2;
}
/* start of main */
int main()
{
 double x1, x2, x3;
 int count;
do
{
 printf("Enter xleft and xright separated by space =");
 scanf("%lf %lf", &x1, &x2);
}
while (f(x1)*f(x2)>0);
/* bisection start */
for (count=0;count< N; count++)
 {
 x3= (x1+x2)/2.0;
 if (f(x1)*f(x3)<0 ) x2=x3; else x1=x3;
 if ( f(x3)==0.0 || fabs(x1-x2)< EPS ) break;
}
 printf("iteration = %d\n", count);
 printf("x= %lf\n", x1);
 return 0;
}
```

The program prompts the user to enter two points (x_1 and x_2) and if $f(x_1)f(x_2) > 0$, the user is prompted again until $f(x_1)f(x_2) < 0$ is satisfied. Note the usage of the `do{...} while{...}` statement. The function, `fabs()`, available in `<math.h>` returns the absolute value of the argument. The output looks like

```
$ gcc bisection.c -lm
$ ./a.out
Enter xleft and xright separated by space =0 1
Enter xleft and xright separated by space =0 2
iteration = 34
x= 1.414214
```

Typically, the bisection method converges after 30–40 iterations for most equations. To provide x_1 and x_2 as an initial guess, it is important to draw a rough graph of $f(x)$ with a graphical application such as gnuplot (see Appendix A) to estimate an approximate interval in which a root is located. Although the bisection method is not the fastest method available, it guaranteers that at least one root can be obtained if the initial interval is chosen correctly.

4.2 NEWTON'S METHOD

4.2.1 NEWTON'S METHOD FOR A SINGLE EQUATION

Newton's method, also known as the Newton-Raphson method, is the de-facto standard of the algorithm to obtain roots for $f(x) = 0$. The convergence of Newton's method is of the second order and instead of specifying two points as in the bisection method, Newton's method requires only one point as initial guess to start with.

As shown in Figure 4.3, Newton's method begins with the initial guess of x_1 which is chosen as close as possible to one of the roots for $f(x) = 0$. At $(x_1, f(x_1))$, a tangent line is drawn as an approximation to $f(x)$ and its intercept point, x_2, with the x axis is deemed as the second approximation which is supposed to be closer to the root. This iteration is repeated until sufficient convergence is attained.

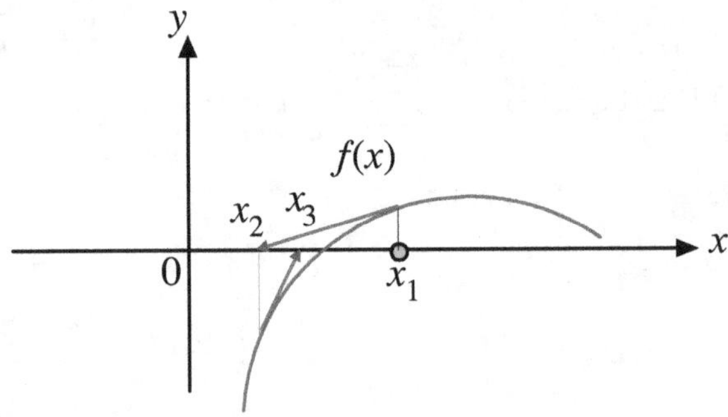

Figure 4.3: Iteration scheme in Newton's method.

From Figure 4.4, the equation of a straight line that passes (a, b) with a slope of m is expressed as

$$y - b = m(x - a).$$

Therefore, the tangent line that passes $(x_1, f(x_1))$ with the slope of $f'(x_1)$ is expressed as

$$y - f(x_1) = f'(x_1)(x - x_1).$$

The condition that this line intercepts with the x axis is

$$0 - f(x_1) = f'(x_1)(x - x_1),$$

which can be solved for x as

$$x = x_1 - \frac{f(x_1)}{f'(x_1)}.$$

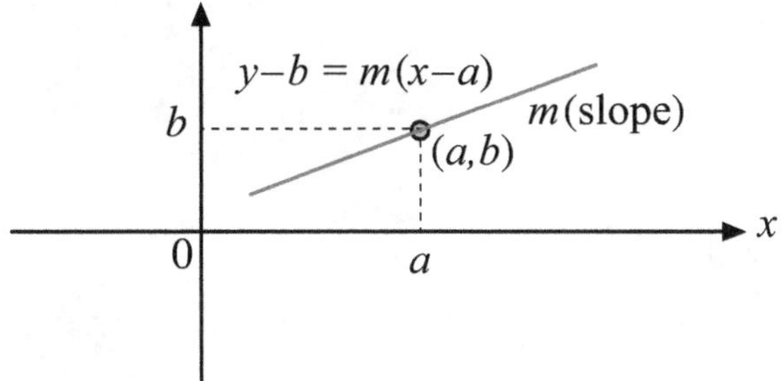

Figure 4.4: Tangent line as approximation to $f(x)$.

Hence, the second approximation is expressed as

$$x_2 = x_1 - \frac{f(x_1)}{f'(x_1)}.$$

In general,

$$x_{n+1} = x_n - \frac{f(x_n)}{f'(x_n)}, \tag{4.1}$$

can be used as as an iterative formula to obtain the $n+1$-th approximation from the n-th approximation. Start with an initial guess of x_1 which is sufficiently close to the root, compute $x_2, x_3, x_4, \ldots$ from Eq. (4.1) and repeat this iteration until $|x_{n+1} - x_n|$ is smaller than the set threshold.

For example, $\sqrt{2}$ can be approximated by solving $f(x) \equiv x^2 - 2 = 0$. With

$$f(x) = x^2 - 2, \quad f'(x) = 2x,$$

Eq. (4.1) is expressed as

$$x_{n+1} = x_n - \frac{x_n^2 - 2}{2x_n}.$$

Therefore, start with $x_1 = 2.0$ (initial guess), then,

$$\begin{aligned} x_1 &= 2, \\ x_2 &= 2 - \frac{f(2)}{f'(2)} = 2 - \frac{2}{4} = 1.5, \\ x_3 &= 1.5 - \frac{f(1.5)}{f'(1.5)} = 1.5 - \frac{0.25}{3} = 1.41667, \\ x_4 &= 1.4167 - \frac{f(1.4167)}{f'(1.4167)} = \ldots = 1.4142. \end{aligned}$$

As is seen above, the convergence is attained after only four iterations.

Here is the algorithm for Newton's method:

1. Pick initial guess, x_1.
2. Confirm that $f'(x_1) \neq 0$.
3. Repeat
   (a) $x_2 \leftarrow x_1 - f(x_1)/f'(x_1)$
   (b) $x_1 \leftarrow x_2$
4. Until $|x_1 - x_2| \leq \epsilon$.

Note that Newton's method fails when $f'(x_n)$ is zero which makes the denominator in Eq. (4.1) zero.

A C program for Newton's method is shown below:

```
#include <stdio.h>
#include <math.h>
#define EPS 1.0e-10
double f(double x)
{
 return x*x-2;
}
double fp(double x)
{
 return 2*x;
}
double newton(double x)
{
 return x - f(x)/fp(x);
}
int main()
{
 double x1, x2;
 int i;
 do{printf("Enter initial guess  =");
  scanf("%lf", &x1);}

 while (fp(x1)==0.0) ;
 for (i=0;i<100;i++)
  {
   x2=newton(x1);
```

```
    if (fabs(x1-x2)< EPS) break;
    x1=x2;
    }
  printf("iteration = %d\n", i);
  printf("x= %lf\n", x1);
  return 0;
}
```

The output looks like

```
$ gcc newton.c -lm
$ ./a.out
Enter initial guess  =2.9
iteration = 5
x= 1.414214
$ ./a.out
Enter initial guess  =0
Enter initial guess  =1.2
iteration = 4
x= 1.414214
```

As is seen in this example, the number of iterations until convergence is 4, much less than the bisection method.

As a side note, the square root of a $(= \sqrt{a})$ can be approximated by solving $f(x) = x^2 - a$. Using Newton's method, the following relation can be derived:

$$
\begin{aligned}
x_{n+1} &= x_n - \frac{x_n^2 - a}{2x_n} \\
&= \frac{1}{2}\left(x_n + \frac{a}{x_n}\right).
\end{aligned}
$$

This iteration scheme can be used to obtain $\sqrt{a}$ even by hand. For example, for $\sqrt{3} = 1.732\ldots$

$$
x_1 = 1,\ x_2 = \frac{1}{2}\left(1 + \frac{3}{1}\right) = 2,\ x_3 = \frac{1}{2}\left(2 + \frac{3}{2}\right) = 1.75,\ x_4 = \frac{1}{2}\left(1.75 + \frac{3}{1.75}\right) = 1.732.
$$

In summary, in the bisection method, convergence is guaranteed with a proper selection of the initial interval but convergence is slow. On the other hand, in Newton's method, convergence is much faster than the bisection method with a proper selection of initial guess.

4.2.2 NEWTON'S METHOD FOR SIMULTANEOUS EQUATIONS (OPTIONAL)

Newton's method can be also used for solving simultaneous equations numerically.[2] For simplicity, two simultaneous equations are given as

$$\begin{aligned} f(x, y) &= 0, \\ g(x, y) &= 0. \end{aligned}$$

By expanding each of the above by the Taylor series for two variable functions, one obtains

$$f(x, y) \sim f(x_0, y_0) + \frac{\partial f}{\partial x}\Big|_{(x_0,y_0)}(x - x_0) + \frac{\partial f}{\partial y}\Big|_{(x_0,y_0)}(y - y_0), \tag{4.2}$$

$$g(x, y) \sim g(x_0, y_0) + \frac{\partial g}{\partial x}\Big|_{(x_0,y_0)}(x - x_0) + \frac{\partial g}{\partial y}\Big|_{(x_0,y_0)}(y - y_0). \tag{4.3}$$

If (x, y) satisfies

$$f(x, y) = 0, \quad g(x, y) = 0,$$

Equations (4.2)–(4.3) can be written as

$$\begin{pmatrix} 0 \\ 0 \end{pmatrix} = \begin{pmatrix} f(x_0, y_0) \\ g(x_0, y_0) \end{pmatrix} + \begin{pmatrix} \frac{\partial f}{\partial x}, & \frac{\partial f}{\partial y} \\ \frac{\partial g}{\partial x}, & \frac{\partial g}{\partial y} \end{pmatrix}\Bigg|_{(x_0,y_0)} \begin{pmatrix} x - x_0 \\ y - y_0 \end{pmatrix},$$

or in vector-matrix form as

$$\mathbf{0} = \mathbf{f}(x_0, y_0) + J(\mathbf{x} - \mathbf{x_o}), \tag{4.4}$$

where

$$J \equiv \begin{pmatrix} \frac{\partial f}{\partial x}, & \frac{\partial f}{\partial y} \\ \frac{\partial g}{\partial x}, & \frac{\partial g}{\partial y} \end{pmatrix}\Bigg|_{(x_0,y_0)}, \quad \mathbf{f}(x_0, y_0) = \begin{pmatrix} f(x_0, y_0) \\ g(x_0, y_0) \end{pmatrix}, \quad \mathbf{x} - \mathbf{x_o} = \begin{pmatrix} x - x_0 \\ y - y_0 \end{pmatrix}.$$

The matrix, J, is called the Jacobian matrix. Equation (4.4) can be solved for $\mathbf{x}$ as

$$\mathbf{x} = \mathbf{x_o} - J^{-1}\mathbf{f}(\mathbf{x_o}), \tag{4.5}$$

where J^{-1} is the inverse matrix of J. Equation (4.5) is the 2-D version of Eq. (4.1).

Example

Numerically solve the following two simultaneous equations for (x, y):

$$x^3 + y^2 = 1, \quad xy = \frac{1}{2}.$$

[2] This topic can be skipped.

(Solution) Let

$$f \equiv x^3 + y^2 - 1, \quad g \equiv xy - \frac{1}{2}.$$

It follows

$$J = \begin{pmatrix} 3x^2, & 2y \\ y, & x \end{pmatrix},$$

and

$$J^{-1} = \begin{pmatrix} \frac{x}{3x^3-2y^2}, & -\frac{2y}{3x^3-2y^2} \\ -\frac{y}{3x^3-2y^2}, & \frac{3x^2}{3x^3-2y^2} \end{pmatrix},$$

therefore,

$$J^{-1}\mathbf{f} = \begin{pmatrix} \frac{x}{3x^3-2y^2}, & -\frac{2y}{3x^3-2y^2} \\ -\frac{y}{3x^3-2y^2}, & \frac{3x^2}{3x^3-2y^2} \end{pmatrix} \begin{pmatrix} x^3 + y^2 - 1 \\ xy - 1/2 \end{pmatrix} = \begin{pmatrix} \frac{x^4 - x(y^2+1)+y}{3x^3-2y^2} \\ \frac{4x^3y - 3x^2 - 2y^3 + 2y}{6x^3-4y^2} \end{pmatrix}.$$

Hence, the iteration scheme is expressed as

$$\begin{pmatrix} x_{n+1} \\ y_{n+1} \end{pmatrix} = \begin{pmatrix} x_n \\ y_n \end{pmatrix} - \begin{pmatrix} \frac{x_n^4 - x_n(y_n^2+1)+y_n}{3x_n^3-2y_n^2} \\ \frac{4x_n^3y_n - 3x_n^2 - 2y_n^3 + 2y_n}{6x_n^3-4y_n^2} \end{pmatrix}. \tag{4.6}$$

Equation (4.6) can be implemented in C as

```
#include <stdio.h>
#include <math.h>
int main()
{
 double x=1.0, y=1.0;
 int i,n;
 printf("Enter x, y and # of iterations=");
 scanf("%lf %lf %d", &x, &y, &n);
 for (i=0;i<=n;i++)
  {
   x = x -(pow(x,4) + y - x*(1 + y*y))/(3*pow(x,3) - 2*y*y);
   y = y -(-3*x*x + 2*y + 4*pow(x,3)*y - 2*pow(y,3))/
                            (6*pow(x,3) - 4*y*y);
  }
 printf("%lf %lf\n", x, y);
 return 0;
}
```

The output looks like

```
$ gcc newton2.c -lm
$ ./a.out
Enter x, y and # of iterations=1 1 4
0.877275 0.569947
$ ./a.out
Enter x, y and # of iterations=1 1 5
0.877275 0.569947
```

Starting $(x, y) = (1.0, 1.0)$, convergence was reached at $(x, y) = (0.877275, 0.569947)$ after only 4 iterations. Note that this is just one of multiple roots. It is necessary to try different initial guess to obtain other roots.

4.2.3 EXERCISES

1. Find all the roots for

$$e^x - 3x = 0,$$

   using the bisection method.

2. Find x for

$$x \sin x = e^x - x \sin\left(x^2\right),$$

   by Newton's method in the interval $[-2, 2]$.

CHAPTER 5

Numerical Differentiation

5.1 INTRODUCTION

It is always possible to analytically differentiate a given function no matter how lengthy or complicated it may be as long as the function is given explicitly. Computer algebra systems such as *Mathematica*[1] and Wolfram Alpha[2] can also differentiate any analytical function exactly.

The only occasion in which numerical differentiation is needed is when a function is given numerically. Consider Table 5.1 that defines $f(x)$ numerically.

Table 5.1: Example of numerically given function

x	$f(x)$
1.0	1.0
1.5	3.375
2.0	8.0
2.5	15.625

The graph of $f(x)$ is shown in Figure 5.1. Our goal is to find $f'(x)$ for each x in the table in terms of difference. Graphically, differentiation is equivalent to the slope (the rate of change). For example, in order to approximate $f'(2.0)$, if one compares $f(2.0)$ with $f(2.5)$, the rate of change is $(f(2.5) - f(2.0))/0.5) = 15.25$ and if one compares $f(2.0)$ with $f(1.5)$, the rate of change is $(f(2.0) - f(1.5))/0.5) = 9.25$. The difference between the two approximations is large and neither one can serve as a good approximation to $f'(2.0)$.

5.2 FORWARD/BACKWARD/CENTRAL DIFFERENCE

There are three basic schemes for numerical differentiation using three neighboring points, all of which can be derived from the Taylor series of $f(x + h)$.

- Forward difference

[1]Wolfram, *The Mathematica Book*, Version 4, Cambridge University Press, 1999.
[2]www.wolframalpha.com

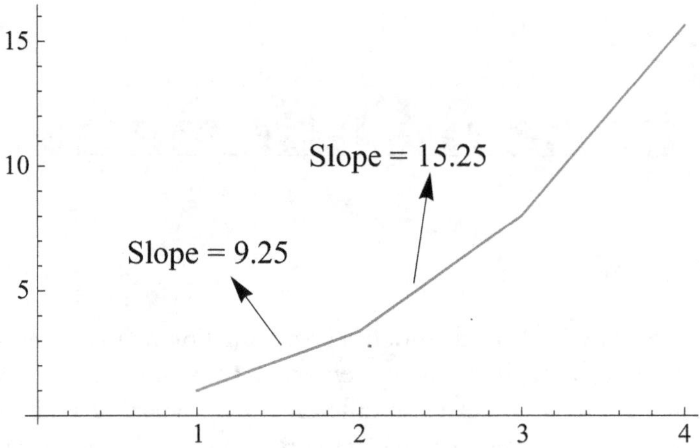

Figure 5.1: Graph of $f(x)$.

In the forward difference scheme, differentiation of $f(x)$ is approximated by comparing $f(x+h)$ and $f(x)$. The Taylor series of $f(x)$ is expressed as

$$\begin{aligned} f(x+h) &= f(x) + hf'(x) + \frac{h^2}{2!}f''(x) + \frac{h^3}{3!}f'''(x) + \ldots \\ &\approx f(x) + hf'(x). \end{aligned} \tag{5.1}$$

By retaining the first two terms of the right-hand side of Eq. (5.1), $f'(x)$ can be expressed as

$$f'(x) \approx \frac{f(x+h) - f(x)}{h}. \tag{5.2}$$

This is called the *forward difference scheme*. Using Eq. (5.2), $f'(2)$ in Table 5.1 is approximated as $(f(2.5) - f(2.0))/0.5) = 15.25$.

- Backward difference

  The backward difference scheme can be obtained by replacing h in Eq. (5.1) by $-h$ as

$$\begin{aligned} f(x-h) &= f(x) - hf'(x) + \frac{h^2}{2!}f''(x) - \frac{h^3}{3!}f'''(x) + \ldots \\ &\approx f(x) - hf'(x), \end{aligned} \tag{5.3}$$

  From Eq. (5.3), $f'(x)$ can be expressed as

$$f'(x) \approx \frac{f(x) - f(x-h)}{h}. \tag{5.4}$$

  This is called the *backward difference scheme*. Using Eq. (5.4), $f'(2)$ in Table 5.1 is approximated as $(f(2.0) - f(1.5))/0.5) = 9.25$.

- Central difference

  Equations (5.1) and (5.3) are listed again as

$$f(x+h) = f(x) + hf'(x) + \frac{h^2}{2!}f''(x) + \frac{h^3}{3!}f'''(x) + \ldots \tag{5.5}$$

$$f(x-h) = f(x) - hf'(x) + \frac{h^2}{2!}f''(x) - \frac{h^3}{3!}f'''(x) + \ldots \tag{5.6}$$

  Subtracting Eq. (5.6) from Eq. (5.5) yields

$$f(x+h) - f(x-h) = 2hf'(x) + 2\frac{h^3}{3!}f'''(x) + \ldots$$

  Dropping the second term and beyond, $f'(x)$ can be expressed as

$$f'(x) \approx \frac{f(x+h) - f(x-h)}{2h}. \tag{5.7}$$

  Equation (5.7) is called the *central difference scheme.*

  By using Eq. (5.7), $f'(2)$ in Table 5.1 is approximated as

$$f'(2) \sim \frac{f(2.5) - f(1.5)}{2 \times 0.5} = 12.25.$$

Revealing that $f(x)$ used in Table 5.1 is $f(x) = x^3$, it follows that $f'(x) = 3x^2$ and hence $f'(2) = 3 \times 2^2 = 12$. It is seen from the above that the central difference yields the best approximation as the truncation error is in the order of h^2 while the truncation error for the forward and backward difference methods is in the order of h.

If the second-order derivative of $f(x)$[3] is desired, adding Eqs. (5.5) and (5.6) yields

$$f(x+h) + f(x-h) = 2f(x) + h^2 f''(x) + \ldots$$

from which $f''(x)$ can be approximated as

$$f''(x) \approx \frac{f(x+h) + f(x-h) - 2f(x)}{h^2}. \tag{5.8}$$

Equation (5.8) can be used to approximate the second-order derivative of $f(x)$.

EXAMPLE

Table 5.2 shows numerical values for $f(x)$ (numerical values of $\sin x$ from $x = 0.0$–1.0).

The following code implements the central difference scheme:

[3]If $f(x)$ represents a position at time, x, $f''(x)$ is its acceleration.

Table 5.2: Example of numerical differentiation

Time	0.0	0.1	0.2	0.3	0.4	0.5
$f(x)$	0	0.0998	0.1986	0.2955	0.3894	0.4794

Time	0.6	0.7	0.8	0.9	1.0
$f(x)$	0.5646	0.6442	0.7173	0.7833	0.8414

```
#include <stdio.h>
#define N 11
int main()
{
 double y[N]={0, 0.0998, 0.1986, 0.2955, 0.3894, 0.4794, 0.5646,
  0.6442, 0.7173, 0.7833, 0.8414};
 double central[N], h=0.1;
 int i;
 for (i=1;i<N-1;i++) central[i]= (y[i+1]-y[i-1])/(2*h);
 printf ("    x    Central \n---------------------------\n");
 for (i=1;i<N-1;i++)  printf ("%lf %lf\n", i*h, central[i]);
 return 0;
}
```

The output is

```
$ gcc central.c
$ ./a.out
    x    Central
---------------------------
0.100000 0.993000
0.200000 0.978500
0.300000 0.954000
0.400000 0.919500
0.500000 0.876000
0.600000 0.824000
0.700000 0.763500
0.800000 0.695500
0.900000 0.620500
```

The central difference scheme is more accurate than the forward difference scheme and the backward difference scheme. However, as seen in the output above, the central difference scheme cannot compute $f'(0.0)$ and $f'(1.0)$ as $f(-0.1)$ and $f(1.1)$ are not available. Using the forward difference scheme for $f'(0.1)$ or the backward difference scheme for $f(1.0)$ is poor compromise.

There is a way to approximate $f'(0.0)$ and $f'(1.0)$ with the same accuracy as the central difference scheme. Replacing h in Eq. (5.6) by $2h$ yields

$$f(x-2h) = f(x) - 2hf'(x) + \frac{4h^2}{2!}f''(x) + \dots \tag{5.9}$$

The h^2 term in Eq. (5.9) can be eliminated by subtracting Eq. (5.9) from 4 times of Eq. (5.6) as

$$4f(x-h) - f(x-2h) = 3f(x) - 2hf'(x) + \text{(higher-order terms)}\dots$$

from which $f'(x)$ can be solved as

$$f'(x) \sim \frac{3f(x) - 4f(x-h) + f(x-2h)}{2h}. \tag{5.10}$$

Equation (5.10) has the same order of accuracy as the central difference scheme. The price that needs to be paid is that instead of two values of $f(x)$, three values must be used. For $x = 1.0$, $f'(1.0)$ can be approximated using $f(1.0)$, $f(0.9)$, and $f(0.8)$. Thus, the previous C code can be modified as

```
#include <stdio.h>
#define N 11
int main()
{
double y[N]={0, 0.0998, 0.1986, 0.2955, 0.3894, 0.4794, 0.5646,
 0.6442, 0.7173, 0.7833, 0.8414};
double central[N], h=0.1;
 int i;
 for (i=1;i<N-1;i++) central[i]= (y[i+1]-y[i-1])/(2*h);
 central[10]=(3*y[10]-4*y[9]+y[8])/(2*h);
 printf ("   x    Central \n---------------------------\n");
 for (i=1;i<N;i++)
 printf ("%lf %lf\n", i*h, central[i]);
 return 0;
}
```

The output is

```
$ gcc prog.c
$ ./a.out
  x     Central
----------------------------
0.100000 0.993000
0.200000 0.978500
0.300000 0.954000
0.400000 0.919500
0.500000 0.876000
0.600000 0.824000
0.700000 0.763500
0.800000 0.695500
0.900000 0.620500
1.000000 0.541500
```

5.3 EXERCISES

1. Derive the formula similar to Eq. (5.10) but at $x = 0$ (first point).

2. The altitude (ft) from the sea level and the corresponding time (sec) for a fictitious rocket were measured as shown in Table 5.3.

Table 5.3: Time and altitude

Time	0	20	40	60	80	100
Altitude	370	9,170	23,835	45,624	62,065	87,368

Time	120	140	160	180	200
Altitude	97,355	103,422	127,892	149,626	160,095

Numerically compute the velocity from the table above for $t = 0$–200 keeping the same accuracy as the central difference scheme. At $t = 200$, use Eq. (5.10) and at $t = 0$, use the formula derived in Problem 1.

CHAPTER 6

Numerical Integration

6.1 INTRODUCTION

Any function if given explicitly can be differentiated analytically but very few functions can be integrated analytically[1] which is why numerical integration is more important than numerical differentiation. While analytical integration is a difficult task in general, it is graphically equivalent to computing the area surrounded by $f(x)$, the x axis and the lower and upper bounds of the integration.

There are three methods widely used to numerically integrate a function, $f(x)$. They are (1) the rectangular rule, (2) the trapezoidal rule, and (3) Simpson's rule. As will be shown in the subsequent sections, Simpson's rule gives the best approximation and is used as the de-facto numerical integration standard.

6.2 RECTANGULAR RULE

As shown in Figure 6.1, the rectangular rule is to approximate $\int_a^b f(x)dx$ by summing up n rectangles over $[a, b]$. Depending on which side of the rectangle to be chosen as the value of $f(x)$, the left rectangular rule and the right rectangular rule can be considered. In the left rectangular

[1]Even a simple function such as $\sin \frac{1}{x}$ cannot be integrated analytically in terms of elementary functions.

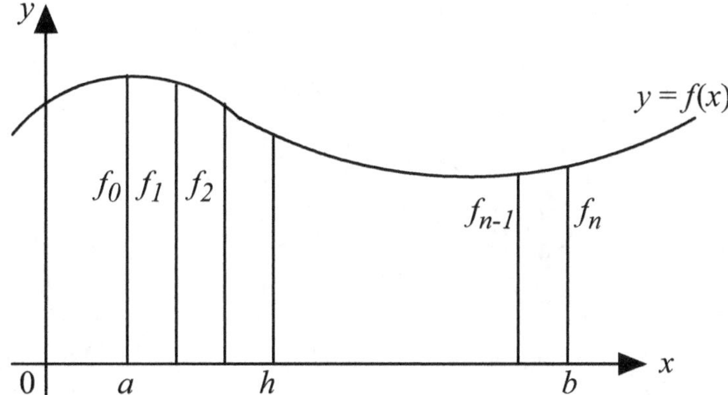

Figure 6.1: Rectangular rule.

rule, the integration, I, can be approximated by

$$\begin{aligned} I &\sim h \times f_0 + h \times f_1 + h \times f_2 + \cdots + h \times f_{n-1} \\ &= h\left(f_0 + f_1 + \cdots + f_{n-1}\right), \end{aligned} \tag{6.1}$$

where $h = (b - a)/n$ is the step size and $f_0, f_1, f_2, \ldots f_{n-1}$ are the selected values of the rectangles equally divided over $[a, b]$ starting from a (the lower bound). As an example, consider

$$I = \int_0^1 \frac{1}{1 + x^2} dx.$$

As this integration can be carried out analytically[2] and the exact value is $\pi/4$, it can be used as a benchmark for the accuracy of each numerical integration scheme. The rectangular rule of Eq. (6.1) can be implemented as

```
#include <stdio.h>
double f(double x)
 {return 4.0/(1.0+x*x) ; }
int main()
{
 int i, n;
 double a=0.0, b=1.0, h, s=0.0 , x ;
 printf("Number of partitions = ");
 scanf("%d", &n) ;
 h = (b-a)/n ;
 for (i= 0;i<n;i++) s = s + f(a + i*h) ;
  s=s*h ;
 printf("Result =%lf\n", s) ;
 return 0;
}
```

The output is

```
$ gcc rectangle.c
$ ./a.out
Number of partitions = 10
Result =3.239926
$ ./a.out
Number of partitions = 100
```

[2] $\int_a^b f(x)dx = \arctan x|_0^1 = \frac{\pi}{4}$.

```
Result =3.151576
$ ./a.out
Number of partitions = 10000
Result =3.141693
$ ./a.out
Number of partitions = 100000
Result =3.141603
$ ./a.out
Number of partitions = 1000000
Result =3.141594
```

It is seen that the convergence is marginal. To obtain accuracy of 5 significant figures, 1,000,000 iterations are needed.

6.3 TRAPEZOIDAL RULE

As shown in Figure 6.2, the trapezoidal rule approximates the integration, I, by a set of trapezoids over the interval.

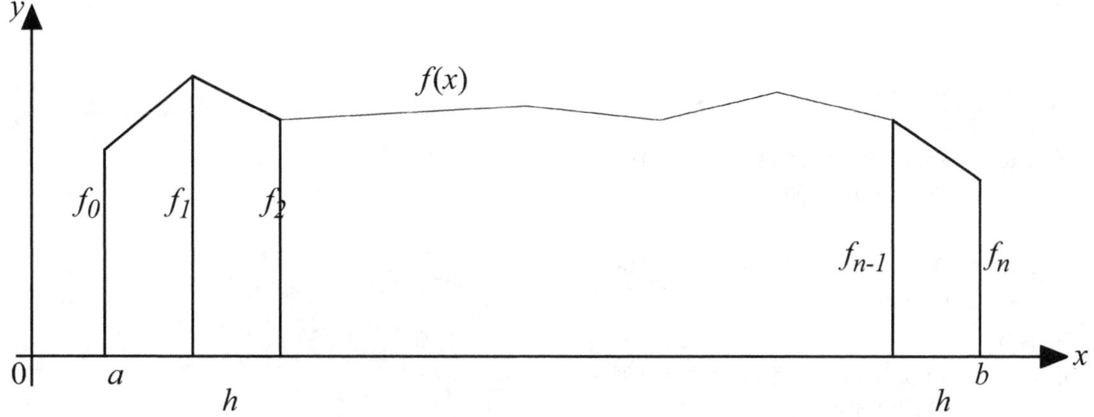

Figure 6.2: Trapezoidal rule.

The sum of all the trapezoids is expressed as

$$\begin{aligned} I &\sim \frac{h}{2}(f_0 + f_1) + \frac{h}{2}(f_1 + f_2) + \cdots + \frac{h}{2}(f_{n-1} + f_n) \\ &= \frac{h}{2}(f_0 + 2f_1 + 2f_2 + \cdots + 2f_{n-1} + f_n) \\ &= \frac{h}{2}(f_0 + f_n) + h \times (f_1 + f_2 + f_3 + \cdots + f_{n-1}). \end{aligned} \tag{6.2}$$

The following code is the implementation of Eq. (6.2):

```
#include <stdio.h>
double f(double x)
 {return 4.0/(1.0+x*x);}
int main()
{
 int i, n ;
 double a=0.0, b=1.0 , h, s=0.0, x;
 printf("Enter number of partitions = ");
 scanf("%d", &n) ;
 h = (b-a)/n ;
 for (i=1;i<=n-1;i++) s = s + f(a + i*h);
 s=h/2*(f(a)+f(b))+ h* s;
 printf("%20.12f\n", s) ;
 return 0;
}
```

The output is

```
$ gcc trapezoid.c
$ ./a.out
Enter number of partitions = 10
      3.139925988907
$ ./a.out
Enter number of partitions = 100
      3.141575986923
$ ./a.out
Enter number of partitions = 1000
      3.141592486923
```

It is seen that the convergence is much faster than the rectangular rule.

6.4 SIMPSON'S RULE

In the rectangular rule, a segment of $f(x)$ is approximated by a flat line (a 0-th order polynomial) and in the trapezoidal rule, it is approximated by a straight line with a slope (a first-order polynomial). As is expected, the next level of refinement is to use a curved line (a second-order polynomial) which leads to Simpson's rule.

First, following Figure 6.3, a second-order polynomial that passes the three points in the figure is sought.

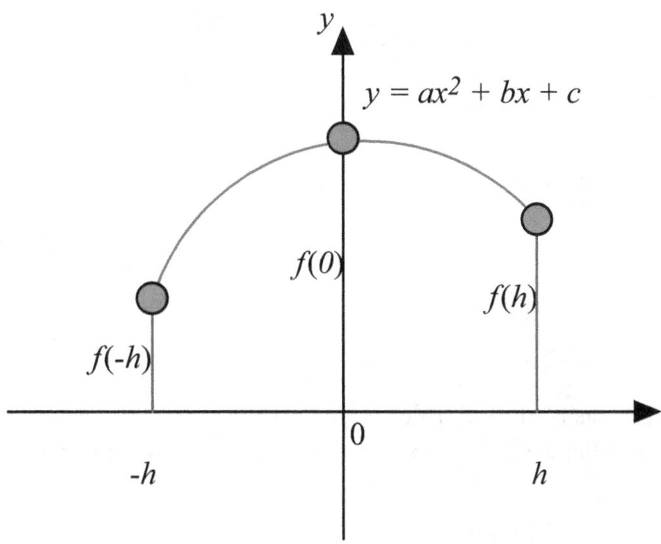

Figure 6.3: Second-order polynomial that passes three points.

The curve that passes $(-h, f(-h))$, $(0, f(0))$ and $(h, f(h))$ is assumed to be

$$y = ax^2 + bx + c.$$

Imposing the condition that this equation passes the three points yields

$$\begin{aligned} f(-h) &= ah^2 - bh + c, \\ f(0) &= c, \\ f(h) &= ah^2 + bh + c, \end{aligned}$$

from which a, b, and c can be solved as

$$\begin{aligned} a &= \frac{f(h) + f(-h) - 2f(0)}{2h^2}, \\ b &= \frac{f(h) - f(-h)}{2h}, \\ c &= f(0). \end{aligned}$$

Now that y is determined, its integral from $-h$ to h is

$$\begin{aligned} \int_{-h}^{h} (ax^2 + bx + c)dx &= \int_{-h}^{h} \left(\frac{f(h) + f(-h) - 2f(0)}{2h^2} x^2 + \frac{f(h) - f(-h)}{2h} x + f(0) \right) dx \\ &= \frac{h}{3} \left(f(-h) + 4f(0) + f(h) \right). \end{aligned} \tag{6.3}$$

The result in Eq. (6.3) implies that the area defined by a curve that passes $(-h, f(-h))$, $(0, f(0))$, and $(h, f(h))$ is expressed as a weighted average of $f(-h)$, $f(0)$ and $f(h)$ with the weight of 1, 4, 1 as

$$\int_{-h}^{h} y\, dx = \frac{f(-h) + 4f(0) + f(h)}{6} \times (2h).$$

By applying this for each segment in the interval $[a, b]$, one obtains

$$\begin{aligned} I &\sim \frac{h}{3}((f_0 + 4f_1 + f_2) + (f_2 + 4f_3 + f_4) + \cdots + (f_{2n-2} + 4f_{2n-1} + f_{2n})) \\ &\sim \frac{h}{3}(f_0 + 4f_1 + 2f_2 + 4f_3 + 2f_4 + \cdots + 2f_{2n-2} + 4f_{2n-1} + f_{2n}), \end{aligned} \tag{6.4}$$

where

$$h = \frac{b-a}{2n}.$$

Equation (6.4) is known as Simpson's rule. Note that the number of partitioning for Simpson's rule must be an even number ($= 2n$). For the coding purpose, it is convenient to rewrite Eq. (6.4) as

$$\begin{aligned} I &= \frac{h}{3}(f_0 + f_{2n}) \\ &+ \frac{h}{3} \times 4(f_1 + f_3 + f_5 + \cdots + f_{2n-1}) \\ &+ \frac{h}{3} \times 2(f_2 + f_4 + f_6 + \cdots + f_{2n-2}). \end{aligned} \tag{6.5}$$

Simpson's rule in Eq. (6.5) can be implemented by the following code:

```
#include <stdio.h>
double f(double x)
 {return 4.0/(1.0+x*x);}
int main()
{
 int i, n ;
 double a=0.0, b=1.0 , h, s1=0.0, s2=0.0, s3=0.0, x;
 printf("Enter number of partitions (must be even) = ");
 scanf("%d", &n) ;
  h = (b-a)/(2.0*n) ;
  s1 = (f(a)+ f(b));
 for (i=1; i<2*n; i=i+2) s2 = s2 + f(a + i*h);
 for (i=2; i<2*n; i=i+2) s3 = s3 + f(a + i*h);
    printf("%20.12lf\n", (h/3.0)*(s1+ 4.0*s2 + 2.0*s3)) ;
```

```
  return 0;
}
```

The output is

```
$ gcc simpson.c
$ ./a.out
Enter number of partitions (must be even) = 10
      3.141592652970
$ ./a.out
Enter number of partitions (must be even) = 20
      3.141592653580
```

As seen in the output above, the convergence is attained with only 10 partitions. Simpson's rule is the de-facto standard of numerical integration. It is tempting to extend Simpson's rule to approximate $f(x)$ by a third order polynomial and beyond. However, such attempts are in general overkill and do not improve accuracy much.

According to error analysis, the truncation error for each rule is as follows:

- Rectangular rule
$$I \sim A + f'(\xi)h$$
- Trapezoidal rule
$$I \sim A + f''(\xi)h^2$$
- Simpson's rule
$$I \sim A + f'''(\xi)h^3$$

where I is the exact integral value, A is its approximation, h is the step size and ξ is a value within the interval. It is noted from the above that the accuracy of numerical integration not only depends on h but also depends on the derivatives of $f(x)$. For example, if one attempts to use $f(x) = \sqrt{1-x^2}$ to approximate $\pi/4$ as in the example code above, its convergence is extremely slow even with Simpson's rule. The slope of $f(x) = \sqrt{1-x^2}$ tends to infinity as x approaches to 1. Thus, $f'(\xi)$, $f''(\xi)$, and $f'''(\xi)$ become singular, which accounts for the slow convergence of the numerical integration.

6.5 EXERCISES

1. (a) Evaluate analytically
$$\int_1^3 2\sqrt{-x^2+4x-3}dx.$$

(b) Write a C program to numerically integrate the above using both the rectangular rule and Simpson's rule.

2. (a) Evaluate analytically

$$\int_0^1 x \ln x dx.$$

(b) Write a C program to numerically integrate the above using Simpson's rule. Note that $\ln x \to -\infty$ as $x \to 0$. So the challenge is how to handle this seemingly singular point at $x = 0$.

CHAPTER 7

Solving Simultaneous Equations

7.1 INTRODUCTION

In Chapter 4, numerical techniques for a single equation, $f(x) = 0$, were discussed. In this chapter, numerical methods for solving a set of simultaneous equations are explained. Many problems in engineering and science end up with solving a set of linear simultaneous equations after they are discretized and linearized. Solving a set of simultaneous equations is one of the most important subjects in numerical analysis. If you took a linear algebra course, you must have learned Cramer's rule which expresses the solution to linear simultaneous equations using the determinants of matrices. However, Cramer's rule is limited to 2 and 3 simultaneous equations and in real-world applications, it is not unusual that the number of equations can be as large as 1,000,000. For such cases, separate methods must be developed.

A set of linear simultaneous equations can be expressed in matrix-vector format as

$$A\mathbf{x} = \mathbf{c},$$

or

$$\begin{pmatrix} a_{11} & a_{12} & \dots & a_{1n} \\ a_{21} & a_{22} & \dots & a_{2n} \\ \vdots & \vdots & \vdots & \vdots \\ a_{n1} & a_{n2} & \dots & a_{nn} \end{pmatrix} \begin{pmatrix} x_1 \\ x_2 \\ \vdots \\ x_n \end{pmatrix} = \begin{pmatrix} c_1 \\ c_2 \\ \vdots \\ c_n \end{pmatrix},$$

or, equivalently,

$$\begin{cases} a_{11}x_1 + a_{12}x_2 + a_{13}x_3 + \dots + a_{1n}x_n = c_1, \\ a_{21}x_1 + a_{22}x_2 + a_{23}x_3 + \dots + a_{2n}x_n = c_2, \\ a_{31}x_1 + a_{32}x_2 + a_{33}x_3 + \dots + a_{3n}x_n = c_3, \\ \qquad \dots\dots \\ a_{n1}x_1 + a_{n2}x_2 + a_{n3}x_3 + \dots + a_{nn}x_n = c_n. \end{cases} \tag{7.1}$$

The number n represents the number of equations and the size of the matrix, A.

CRAMER'S RULE

For two and three simultaneous equations, Cramer's rule is appropriate. The determinant for a 2×2 matrix is defined as

$$\begin{vmatrix} a_{11} & a_{12} \\ a_{21} & a_{22}. \end{vmatrix} \equiv a_{11}a_{22} - a_{12}a_{21}.$$

Similarly, the determinant for a 3×3 matrix is defined as

$$\begin{vmatrix} a_{11} & a_{12} & a_{13} \\ a_{21} & a_{22} & a_{23} \\ a_{31} & a_{32} & a_{33} \end{vmatrix} \equiv a_{11}a_{22}a_{33} + a_{12}a_{23}a_{31} + a_{21}a_{32}a_{13} - a_{13}a_{22}a_{31} - a_{12}a_{21}a_{33} - a_{23}a_{32}a_{11}.$$

The determinant of $n \times n (n > 3)$ matrices can be expressed similar to the above. It consists of $n!$ terms each of which is a product of n elements.

Using the determinants, the solutions for the following two simultaneous equations

$$\begin{aligned} a_{11}x_1 + a_{12}x_2 &= c_1, \\ a_{21}x_1 + a_{22}x_2 &= c_2, \end{aligned}$$

are expressed as

$$\begin{aligned} x_1 &= \frac{\begin{vmatrix} c_1 & a_{12} \\ c_2 & a_{22} \end{vmatrix}}{\begin{vmatrix} a_{11} & a_{12} \\ a_{21} & a_{22} \end{vmatrix}} = \frac{c_1a_{22} - c_2a_{12}}{a_{11}a_{22} - a_{12}a_{21}}, \\ x_2 &= \frac{\begin{vmatrix} a_{11} & c_1 \\ a_{21} & c_2 \end{vmatrix}}{\begin{vmatrix} a_{11} & a_{12} \\ a_{21} & a_{22} \end{vmatrix}} = \frac{c_2a_{11} - c_1a_{21}}{a_{11}a_{22} - a_{12}a_{21}}. \end{aligned}$$

Similarly, for three simultaneous equations,

$$\begin{aligned} a_{11}x_1 + a_{12}x_2 + a_{13}x_3 &= c_1, \\ a_{21}x_1 + a_{22}x_2 + a_{23}x_3 &= c_2, \\ a_{31}x_1 + a_{32}x_2 + a_{33}x_3 &= c_3, \end{aligned}$$

the solution is expressed as

$$x_1 = \frac{\begin{vmatrix} c_1 & a_{12} & a_{13} \\ c_2 & a_{22} & a_{23} \\ c_3 & a_{32} & a_{33} \end{vmatrix}}{\begin{vmatrix} a_{11} & a_{12} & a_{13} \\ a_{21} & a_{22} & a_{23} \\ a_{31} & a_{32} & a_{33} \end{vmatrix}},$$

$$x_2 = \frac{\begin{vmatrix} a_{11} & c_1 & a_{13} \\ a_{21} & c_2 & a_{23} \\ a_{31} & c_3 & a_{33} \end{vmatrix}}{\begin{vmatrix} a_{11} & a_{12} & a_{13} \\ a_{21} & a_{22} & a_{23} \\ a_{31} & a_{32} & a_{33} \end{vmatrix}},$$

$$x_3 = \frac{\begin{vmatrix} a_{11} & a_{12} & c_1 \\ a_{21} & a_{22} & c_2 \\ a_{31} & a_{32} & c_3 \end{vmatrix}}{\begin{vmatrix} a_{11} & a_{12} & a_{13} \\ a_{21} & a_{22} & a_{23} \\ a_{31} & a_{32} & a_{33} \end{vmatrix}}.$$

The above formulas for the solutions are called *Cramer's rule* and it is possible to extend Cramer's rule for a set of more than three simultaneous equations using the determinant for matrices of larger size. However, the practical usage of Cramer's rule is limited to two and three simultaneous equations. This is due to the efficiency of computing the determinant. In general, the determinant for an $n \times n$ matrix consists of $n!$ terms as shown for 2×2 and 3×3 matrices above. Each term has $n - 1$ multiplications. Therefore, the number of multiplications required for Cramer's rule for n simultaneous equations is $n!(n-1)(n+1)$ counting the denominator. For $n = 4$, this is 360 and for $n = 10$, this amounts to 359,251,200. Approximate computational time for n simultaneous equations with a 100 MFLOPS computer (an old PC) using Cramer's rule is estimated, as shown in Table 7.1.[1]

Table 7.1: Time required for Cramer's rule

n	10	12	14	16	18	20
Time	0.4 sec	1 min	3.6 h	41 days	38 yr	16,000 yr

[1]Dahmen and Reusken, *Numerik für Ingenieure und Naturwissenschaftler*, Springer, 2006.

7.2 GAUSS-JORDAN ELIMINATION METHOD

The Gauss-Jordan elimination method[2] is a practical method to systematically solve a set of many simultaneous equations numerically. It is less efficient than the LU decomposition method discussed in Section 7.3 but is widely taught as one of the primary numerical techniques for simultaneous equations.

It is based on the principle of linearity (also called the principle of superposition) in which if there are two linear equations, their linear combination is also yet another equation. For example, consider the following two equations:

$$a_{11}x_1 + a_{12}x_2 + \ldots + a_{1n}x_n = c_1, \tag{7.2}$$

$$a_{21}x_1 + a_{22}x_2 + \ldots + a_{2n}x_n = c_2. \tag{7.3}$$

Multiplying λ_1 by Eq. (7.2) and λ_2 by Eq. (7.3) and adding the two yield

$$(\lambda_1 a_{11} + \lambda_2 a_{21})x_1 + (\lambda_1 a_{12} + \lambda_2 a_{22})x_2 + \ldots + (\lambda_1 a_{1n} + \lambda_2 a_{2n})x_n = \lambda_1 c_1 + \lambda_2 c_2. \tag{7.4}$$

Equation (7.4) is yet another valid linear equation.

The Gauss-Jordan elimination method is to choose λ appropriately to successively convert Eqs. (7.1) into

$$\begin{cases} 1 \times x_1 + 0 \times x_2 + 0 \times x_3 + \ldots + 0 \times x_n = d_1, \\ 0 \times x_1 + 1 \times x_2 + 0 \times x_3 + \ldots + 0 \times x_n = d_2, \\ 0 \times x_1 + 0 \times x_2 + 1 \times x_3 + \ldots + 0 \times x_n = d_3, \\ \qquad \cdots\cdots \\ 0 \times x_1 + 0 \times x_2 + 0 \times x_3 + \ldots + 1 \times x_n = d_n. \end{cases}$$

The sequence, $\{d_1, d_2, d_3, \ldots d_n\}$, is the solution for $\{x_1, x_2, x_3, \ldots x_n\}$.

EXAMPLE

The Gauss-Jordan elimination method is illustrated by the following example. The goal is to convert Eq. (7.5) to Eq. (7.6) by using the diagonal elements as pivots.

$$\left.\begin{aligned} 2x - y + z &= 2, \\ -x + 3y + 3z &= 3, \\ 2x + y + 4z &= 1. \end{aligned}\right\} \tag{7.5}$$

$$\Rightarrow$$

$$\left.\begin{aligned} 1x + 0y + 0z &= ?, \\ 0x + 1y + 0z &= ?, \\ 0x + 0y + 1z &= ?. \end{aligned}\right\} \tag{7.6}$$

Table 7.2 can be used for this conversion.

[2]Also known as the Gaussian elimination or row reduction method.

Table 7.2: Steps in the Gauss-Jordan elimination method

Reference Line Number	x	y	z	=	Comment
(1)	2	-1	1	2	
(2)	-1	3	3	3	
(3)	2	1	4	1	
(4)	1	-1/2	1/2	1	(1) ÷ 2
(5)	-1	3	3	3	
(6)	2	1	4	1	
(7)	1	-1/2	1/2	1	
(8)	0	5/2	7/2	4	(5) — (4) × (—1)
(9)	0	2	3	-1	(6) — (4) × 2
(10)	1	-1/2	1/2	1	
(11)	0	1	7/5	8/5	(8) ÷ (5/2)
(12)	0	2	3	-1	
(13)	1	0	6/5	9/5	(10) — (11) × (— 1/2)
(14)	0	1	7/5	8/5	
(15)	0	0	1/5	-21/5	(12) — (11) × 2
(16)	1	0	6/5	9/5	
(17)	0	1	7/5	8/5	
(18)	0	0	1	-21	(15) ÷ (1/5)
(19)	1	0	0	27	(16) — (18) × (6/5)
(20)	0	1	0	31	(17) — (18) × (7/5)
(21)	0	0	1	-21	

In the Gauss-Jordan elimination method, the elements in the first row of Table 7.2 are divided by a_{11} so that a_{11} is normalized to be 1 (line (4)). Using $a_{11} = 1$ as a pivot, a_{21} and a_{31} are eliminated (lines (8)–(9)). Next, the elements in the second row (line (8)) are divided by a_{22} so that a new $a_{22} = 1$ is used as the next pivot. Using $a_{22} = 1$, a_{12} and a_{32} are eliminated (lines (13)–(15)). Repeat this for a_{33}. Finally, the lines (19)–(21) are read as

$$x = 27, \quad y = 31, \quad z = -21.$$

It can be shown that the number of multiplications required for the Gauss-Jordan elimination method is approximately proportional to n^3. For $n = 10$, this amounts to 1,000. Compare this number with the one required for Cramer's rule (359,251,200).

The inverse of the matrix A can be obtained using the Gauss-Jordan elimination method by listing the matrix A along with the identity matrix I as

$$(A \mid I),$$

i.e.,

$$\begin{pmatrix} 2 & -1 & 1 & 1 & 0 & 0 \\ -1 & 3 & 3 & 0 & 1 & 0 \\ 2 & 1 & 4 & 0 & 0 & 1 \end{pmatrix} \rightarrow \begin{pmatrix} 1 & -\frac{1}{2} & \frac{1}{2} & \frac{1}{2} & 0 & 0 \\ 0 & \frac{5}{2} & \frac{7}{2} & \frac{1}{2} & 1 & 0 \\ 0 & 2 & 3 & -1 & 0 & 1 \end{pmatrix}$$
$$\rightarrow \begin{pmatrix} 1 & 0 & \frac{6}{5} & \frac{3}{5} & \frac{1}{5} & 0 \\ 0 & 1 & \frac{7}{5} & \frac{1}{5} & \frac{2}{5} & 0 \\ 0 & 0 & \frac{1}{5} & -\frac{7}{5} & -\frac{4}{5} & 1 \end{pmatrix} \rightarrow \begin{pmatrix} 1 & 0 & 0 & 9 & 5 & -6 \\ 0 & 1 & 0 & 10 & 6 & -7 \\ 0 & 0 & 1 & -7 & -4 & 5 \end{pmatrix}.$$

The inverse of A is stored as the second half of the whole matrix as

$$A^{-1} = \begin{pmatrix} 9 & 5 & -6 \\ 10 & 6 & -7 \\ -7 & -4 & 5 \end{pmatrix}.$$

The following code is implementation of the Gauss-Jordan elimination method. It should be noted that the index of an array in C begins at 0 while in linear algebra, the index of vectors and matrices begins with 1. Therefore, when implementing equations in linear algebra in C, every index needs to be shifted to the left by 1. In the following program, the combination of the matrix A and the vector c can be entered into an array, `a[3][4]`.

```
#include <stdio.h>
#define N 3
int main()
{
 double a[N][N+1]={{2, -1, 1, 2},{-1, 3, 3, 3},{2, 1, 4, 1}};
 double pivot,d;
 int i,j,k;

 for(k=0; k<N; k++)
 {
  pivot=a[k][k];

  for(j=k; j<N+1; j++) a[k][j]=a[k][j]/pivot;
  for(i=0; i<N;  i++)
  {
    if(i != k)
```

```
      {
        d=a[i][k];
         for(j=k; j<N+1; j++) a[i][j]=a[i][j]-d*a[k][j];
      }
    }
  }

  for(i=0; i<N; i++) printf("x[%d]=%lf\n", i+1, a[i][N]);
  return 0;
}
```

The output is

```
$ gcc gaussjordan.c
$ ./a.out
x[1]=27.000000
x[2]=31.000000
x[3]=-21.000000
```

7.3 LU DECOMPOSITION (OPTIONAL)

The Gauss-Jordan elimination method is appropriate for solving a large set of linear simultaneous equations.

The LU decomposition[3] (also known as the LU factorization) is a refinement of the Gauss-Jordan elimination method that further reduces the number of operations resulting in faster execution. With the LU decomposition, the number of operations can be reduced to the order of $n^3/3$ compared with n^3 for the Gauss-Jordan elimination method.

First, any matrix, A, can be uniquely factorized as

$$A = LU, \tag{7.7}$$

where L and U are lower and upper triangular matrices whose components are expressed as

$$L = \begin{pmatrix} 1 & 0 & 0 & \ldots & 0 \\ l_{21} & 1 & 0 & \ldots & 0 \\ l_{31} & l_{32} & 1 & \ldots & 0 \\ \vdots & \vdots & \vdots & \ddots & \vdots \\ l_{n1} & l_{n2} & l_{n3} & \ldots & 1 \end{pmatrix}, \quad U = \begin{pmatrix} u_{11} & u_{12} & u_{13} & \ldots & u_{1n} \\ 0 & u_{22} & u_{23} & \ldots & u_{2n} \\ 0 & 0 & u_{33} & \ldots & u_{3n} \\ \vdots & \vdots & \vdots & \ddots & \vdots \\ 0 & 0 & 0 & \ldots & u_{nn} \end{pmatrix}.$$

[3]This topic can be skipped.

Note that the diagonal elements of L are set to be 1. This decomposition is called *LU decomposition* (or *LU factorization*) and provides the most efficient way of solving simultaneous equations.

The decomposition of Eq. (7.7) is unique as it is possible to find L and U directly. For example, for a 4×4 matrix, Eq. (7.7) can be written as

$$\begin{pmatrix} 1 & 0 & 0 & 0 \\ l_{21} & 1 & 0 & 0 \\ l_{31} & l_{32} & 1 & 0 \\ l_{41} & l_{42} & l_{43} & 1 \end{pmatrix} \begin{pmatrix} u_{11} & u_{12} & u_{13} & u_{14} \\ 0 & u_{22} & u_{23} & u_{24} \\ 0 & 0 & u_{33} & u_{34} \\ 0 & 0 & 0 & u_{44} \end{pmatrix} = \begin{pmatrix} a_{11} & a_{12} & a_{13} & a_{14} \\ a_{21} & a_{22} & a_{23} & a_{24} \\ a_{31} & a_{32} & a_{33} & a_{34} \\ a_{41} & a_{42} & a_{43} & a_{44} \end{pmatrix}. \tag{7.8}$$

Writing each element in Eq. (7.8) explicitly gives

$$\begin{cases} u_{11} = a_{11}, & u_{12} = a_{12}, & u_{13} = a_{13}, \\ l_{21}u_{11} = a_{21}, & l_{21}u_{12} + u_{22} = a_{22}, & l_{21}u_{13} + u_{23} = a_{23}, \\ l_{31}u_{11} = a_{31}, & l_{31}u_{12} + l_{32}u_{22} = a_{32}, & l_{31}u_{13} + l_{32}u_{23} + u_{33} = a_{33}, \\ l_{41}u_{11} = a_{41}, & l_{41}u_{12} + l_{42}u_{22} = a_{42}, & l_{41}u_{13} + l_{42}u_{23} + l_{43}u_{33} = a_{43}, \end{cases}$$

$$\begin{aligned} & u_{14} = a_{14}, \\ & l_{21}u_{14} + u_{24} = a_{24}, \\ & l_{31}u_{14} + l_{32}u_{24} + u_{34} = a_{34}, \\ & l_{41}u_{14} + l_{42}u_{24} + l_{43}u_{34} + u_{44} = a_{44}, \end{aligned}$$

from which all the elements of l_{ij} and u_{ij} can be solved as

$$\begin{cases} u_{11} = a_{11}, & u_{12} = a_{12}, & u_{13} = a_{13}, \\ l_{21} = \frac{a_{21}}{u_{11}}, & u_{22} = a_{22} - l_{21}u_{12}, & u_{23} = a_{23} - l_{21}u_{13}, \\ l_{31} = \frac{a_{31}}{u_{11}}, & l_{32} = \frac{a_{32} - l_{31}u_{12}}{u_{22}}, & u_{33} = a_{33} - l_{31}u_{13} - l_{32}u_{23}, \\ l_{41} = \frac{a_{41}}{u_{11}}, & l_{42} = \frac{a_{42} - l_{41}u_{12}}{u_{22}}, & l_{43} = \frac{a_{43} - l_{41}u_{13} - l_{42}u_{23}}{u_{33}}, \end{cases}$$

$$\begin{aligned} & u_{14} = a_{14}, \\ & u_{24} = a_{24} - l_{21}u_{14}, \\ & u_{34} = a_{34} - l_{31}u_{14} - l_{32}u_{24}, \\ & u_{44} = a_{44} - l_{41}u_{14} - l_{42}u_{24} - l_{43}u_{34}. \end{aligned}$$

Note that the immediate results for u_{ij} and l_{ij} are used to solve for the next result.

Using the LU decomposition, $A\mathbf{x} = \mathbf{c}$ is written as

$$LU\mathbf{x} = \mathbf{c}. \tag{7.9}$$

To solve Eq. (7.9) for $\mathbf{x}$, the following two steps are needed:

1. Define $\mathbf{y} \equiv U\mathbf{x}$ and solve $L\mathbf{y} = \mathbf{c}$ for $\mathbf{y}$ first.

2. After $\mathbf{y}$ is solved, solve $U\mathbf{x} = \mathbf{y}$ for $\mathbf{x}$.

It appears first that the two steps above to solve two equations require more computation than solving a single equation of $A\mathbf{x} = \mathbf{c}$. However, finding L and U from A requires much less effort and solving for the two equations is almost trivial.

The first equation, $L\mathbf{y} = \mathbf{c}$, can be written as

$$\begin{pmatrix} 1 & 0 & 0 & 0 \\ l_{21} & 1 & 0 & 0 \\ l_{31} & l_{32} & 1 & 0 \\ l_{41} & l_{42} & l_{43} & 1 \end{pmatrix} \begin{pmatrix} y_1 \\ y_2 \\ y_3 \\ y_4 \end{pmatrix} = \begin{pmatrix} c_1 \\ c_2 \\ c_3 \\ c_4 \end{pmatrix},$$

or written explicitly as

$$\begin{cases} y_1 = c_1, \\ l_{21}y_1 + y_2 = c_2, \\ l_{31}y_1 + l_{32}y_2 + y_3 = c_3, \\ l_{41}y_1 + l_{42}y_2 + l_{43}y_3 + y_4 = c_4. \end{cases}$$

The solution for $y_1 \sim y_4$ is straightforward as

$$\begin{cases} y_1 = c_1, \\ y_2 = c_2 - l_{21}y_1, \\ y_3 = c_3 - l_{31}y_1 - l_{32}y_2, \\ y_4 = c_4 - l_{41}y_1 - l_{42}y_2 - l_{43}y_3. \end{cases}$$

The second equation, $U\mathbf{x} = \mathbf{y}$, can be written as

$$\begin{pmatrix} u_{11} & u_{12} & u_{13} & u_{14} \\ 0 & u_{22} & u_{23} & u_{24} \\ 0 & 0 & u_{33} & u_{34} \\ 0 & 0 & 0 & u_{44} \end{pmatrix} \begin{pmatrix} x_1 \\ x_2 \\ x_3 \\ x_4 \end{pmatrix} = \begin{pmatrix} y_1 \\ y_2 \\ y_3 \\ y_4 \end{pmatrix},$$

or, equivalently, as

$$\begin{cases} u_{11}x_1 + u_{12}x_2 + u_{13}x_3 + u_{14}x_4 = y_1, \\ u_{22}x_2 + u_{23}x_3 + u_{24}x_4 = y_2, \\ u_{33}x_3 + u_{34}x_4 = y_3, \\ u_{44}x_4 = y_4. \end{cases}$$

Solving for $x_1 \sim x_4$ is straightforward as well and yields

$$\begin{cases} x_4 = \frac{y_4}{u_{44}}, \\ x_3 = \frac{y_3 - u_{34}x_4}{u_{33}}, \\ x_2 = \frac{y_2 - u_{23}x_3 - u_{24}x_4}{u_{22}}, \\ x_1 = \frac{y_1 - u_{11}x_1 - u_{12}x_2 - u_{13}x_3}{u_{11}}. \end{cases}$$

This procedure is called backward substitution.

It can be shown that the number of operations (multiplications, divisions) to decompose A into $A = LU$ is about $n^3/3$ and the number to obtain x from $Ly = c$ and $Ux = y$ is proportional to n^2 so the total number of operations is in the order of $n^3/3$[4] which is 1/3 of the number required for the Gauss-Jordan elimination method.

Here is a C code to implement the LU decomposition solving the same equations as the example in the Gauss-Jordan elimination method.

```
#include <stdio.h>
#define N 3
int main()
{
double a[N][N+1]={{2, -1, 1, 2},{-1, 3, 3, 3},{2, 1, 4, 1}};
int i,j,k,l;

/* LU decomposition and forward reduction */
 for(j=1;j<N+1;j++)
  a[0][j] = a[0][j]/a[0][0];
   for(k=1;k<N;k++){
    for(i=k;i<N;i++)
     for(l=0;l<k;l++)
      a[i][k] =a[i][k]- a[i][l]*a[l][k];
       for(j=k+1;j<N+1;j++){
        for(l=0;l<k;l++)
         a[k][j]= a[k][N]-a[k][l]*a[l][j];
         a[k][j]= a[k][j]/a[k][k];
        }
      }

/* Backward substitution */
for(k=N-1;k>=0;--k){
 for(l=k+1;l<N;l++){
  a[k][N] = a[k][N]-a[k][l]*a[l][N];
  }
 }

/* print the result */
for(i=0;i<N;i++){
```

[4] n^2 can be neglected compared with n^3.

```
  printf("x[%d]=%lf",i, a[i][N]); printf("\n");
  }
 return 0;
}
```

The output is

```
$ gcc lu.c
$ ./a.out
x[0]=27.000000
x[1]=31.000000
x[2]=-21.000000
```

NOTES

1. If A is a symmetric matrix ($a_{ij} = a_{ji}$), the LU-decomposition is called the Cholesky decomposition. The number of operations required for the Cholesky decomposition is further reduced to approximately $n^3/6$.

2. If A is a sparse matrix, so are L and U.

3. If A is a triangular diagonal matrix (typical of matrices used in the finite element method), so are L and U.

7.4 GAUSS-SEIDEL METHOD (JACOBI METHOD)

The Gauss-Seidel method and the Jacobi method are iterative methods to solve a set of certain types of simultaneous equations. Unlike the Gauss-Jordan elimination method, there is no guarantee that the iterations ever converge but no thorough programming is required and the method can be also applied to certain nonlinear simultaneous equations.

To illustrate these methods, consider the following three simultaneous equations:

$$\left\{\begin{array}{rcl} 7x + y + 2z & = & 10, \\ x + 8y + 3z & = & 8, \\ 2x + 3y + 9z & = & 6. \end{array}\right. \tag{7.10}$$

Equations (7.10) can be written as

$$\left\{\begin{array}{rcl} x & = & (10 - y - 2z)/7, \\ y & = & (8 - x - 3z)/8, \\ z & = & (6 - 2x - 3y)/9. \end{array}\right. \tag{7.11}$$

Equations (7.11) are rewritten in the form of an iterative scheme as

$$\begin{cases} x_{n+1} &= (10 - y_n - 2z_n)/7, \\ y_{n+1} &= (8 - x_n - 3z_n)/8, \\ z_{n+1} &= (6 - 2x_n - 3y_n)/9. \end{cases} \tag{7.12}$$

Equations (7.12) can be solved iteratively starting with x_0, y_0, z_0 chosen by initial guess. This scheme is called the *Jacobi method*. A slightly better iteration scheme is to use the most immediate preceding values for the next approximation, i.e.,

$$\begin{cases} x_{n+1} &= (10 - y_n - 2z_n)/7, \\ y_{n+1} &= (8 - x_{n+1} - 3z_n)/8, \\ z_{n+1} &= (6 - 2x_{n+1} - 3y_{n+1})/9. \end{cases} \tag{7.13}$$

This scheme is called the *Gauss-Seidel method*. The Gauss-Seidel method is faster than the Jacobi method as it uses the most immediate values for the iterations. It is straightforward to implement the Gauss-Seidel method. As any statement in C uses the most immediate value for variable substitutions, the Gauss-Seidel method is automatically chosen over the Jacobi method when coding. In the following code, $(x_0, y_0, z_0) = (0, 0, 0)$ was chosen as the initial guess:

```
#include <stdio.h>
int main()
{
  double x, y, z;
  int i,n;
  x=y=z=0.0;
  printf("Enter # of iteration = ");
  scanf("%d", &n);

  for (i=0;i<n;i++)
   {
    x = (10-y-2*z)/7;
    y = (8-x-3*z)/8.0;
    z = (6-2*x-3*y)/9.0;
    }
  printf("x = %lf, y= %lf, z=%lf\n", x,y,z);
  return 0;
}
```

The output is

```
$ gcc gauss-seidel.c
$ ./a.out
Enter # of iteration = 9
x = 1.281553, y= 0.796117, z=0.116505
$ ./a.out
Enter # of iteration = 10
x = 1.281553, y= 0.796117, z=0.116505
```

It is seen that the convergence was attained after nine iterations.

There may be occasions when the Jacobi method or the Gauss-Seidel method does not converge. In general, if the diagonal elements of the matrix, A, are larger than the rest of the non-diagonal elements, the Jacobi method or the Gauss-Seidel method works for $A\mathbf{x} = \mathbf{c}$.[5] Problem 2 in Exercises 7.5 is an example of solving a set of nonlinear equations using the Gauss-Seidel method.

7.5 EXERCISES

1. Solve the following five simultaneous equations by the Gauss-Jordan elimination method:

$$\begin{cases} a_{11}x_1 + a_{12}x_2 + a_{13}x_3 + a_{14}x_4 + a_{15}x_5 = c_1, \\ a_{21}x_1 + a_{22}x_2 + a_{23}x_3 + a_{24}x_4 + a_{25}x_5 = c_2, \\ a_{31}x_1 + a_{32}x_2 + a_{33}x_3 + a_{34}x_4 + a_{35}x_5 = c_3, \\ a_{31}x_1 + a_{32}x_2 + a_{33}x_3 + a_{44}x_4 + a_{45}x_5 = c_4, \\ a_{n1}x_1 + a_{n2}x_2 + a_{n3}x_3 + a_{54}x_4 + a_{55}x_5 = c_5, \end{cases}$$

where a_{ij} is given as

```
a[5][5]={
{3.55618, 5.87317, 7.84934, 5.6951, 3.84642},
{-4.82893, 8.38177, -0.301221, 5.10182, -4.1169},
{-7.64196, 5.66605,3.20481, 1.55619, -1.19814},
{-2.95914, -9.16958,7.3216, 2.39876, -8.1302},
{-8.42043, -0.369407, -5.4102, -8.00545, 9.22153}
};
```

[5]To be more exact, if the largest eigenvalue, λ, of A is greater than 1, the scheme converges.

and c_i is given as

```
c[5]={-1.92193, -2.35262, 2.27709, -2.67493, 1.84756};
```

2. Solve the following set of nonlinear equations by the Gauss-Seidel method:

$$\begin{cases} 27x + e^x \cos y - 0.12z & = & 3, \\ -0.2x^2 + 37y + 3xz & = & 6, \\ x^2 - 0.2y \sin x + 29z & = & -4. \end{cases} \tag{7.14}$$

Start with an initial guess of $x = y = z = 1$.

CHAPTER 8

Differential Equations

Differential equations are the most important type of equations in scientific computation as every physical phenomenon in nature is described by differential equations. We can sense physical objects because we can capture the rate of the change of the objects and the equations to describe this rate of change are the differential equations. Solving the Navier-Stokes equations in fluid mechanics can predict weather patterns as weather is determined by the movement of air particles. Similarly, solving the stress equilibrium equations in solid mechanics can predict the strength and durability of materials used and solving the energy equation in heat transfer can predict the temperature distribution around us all of which are described by differential equations.

However, solving differential equations analytically is a daunting task to say the least. Differential equations have been studied over the past 400 years but many useful and important differential equations have no analytical solutions available. The good news is that solving differential equations (initial value problems) numerically does not require an advanced level of mathematics as will be illustrated in this chapter.

There are two types of differential equations—initial value problems and boundary value problems. This chapter restricts the scope to initial value problems only and Euler's method and the Runge-Kutta method are introduced as the numerical schemes.

8.1 INITIAL VALUE PROBLEMS

The basic form of a differential equation for initial value problems can be expressed as

$$\frac{dy}{dt} = f(t, y), \tag{8.1}$$

where y is an unknown function, t is the independent variable, and $f(t, y)$ is a function of t and y. Equation (8.1) along with the initial condition of

$$y(t_0) = y_0$$

is called the initial value problem.

Equation (8.1) can be interpreted graphically as shown in Figure 8.1. Solving for $y(x)$ in Eq. (8.1) is equivalent to drawing a curve starting at (t_0, y_0) whose slope at (t_1, y_1) is given as $f(t_1, y_1)$. Only the starting point and a slope dy/dt, not y, at an arbitrary point are provided.

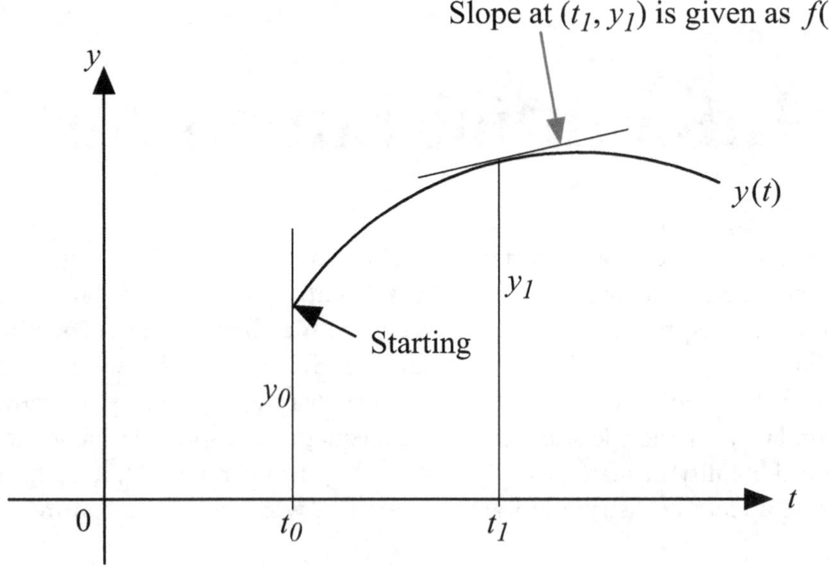

Figure 8.1: Differential equation (initial value problem).

8.1.1 EULER'S METHOD

Euler's method is a primitive method for solving initial value problems numerically.

In Euler's method, dy/dt in Eq. (8.1) is approximated by the forward difference as

$$\frac{dy}{dt} \sim \frac{y_{n+1} - y_n}{h}.$$

Therefore, Eq. (8.1) is approximated as

$$\frac{y_{n+1} - y_n}{h} = f(t, y). \tag{8.2}$$

Equation (8.2) can be written as

$$y_{n+1} = y_n + hf(t, y), \tag{8.3}$$

which can be used to predict y_{n+1} from y_n and the slope at that point as shown in Figure 8.2.

Example 1

Consider the following initial value problem:

$$\frac{dy}{dt} = y, \quad y(0) = 1. \tag{8.4}$$

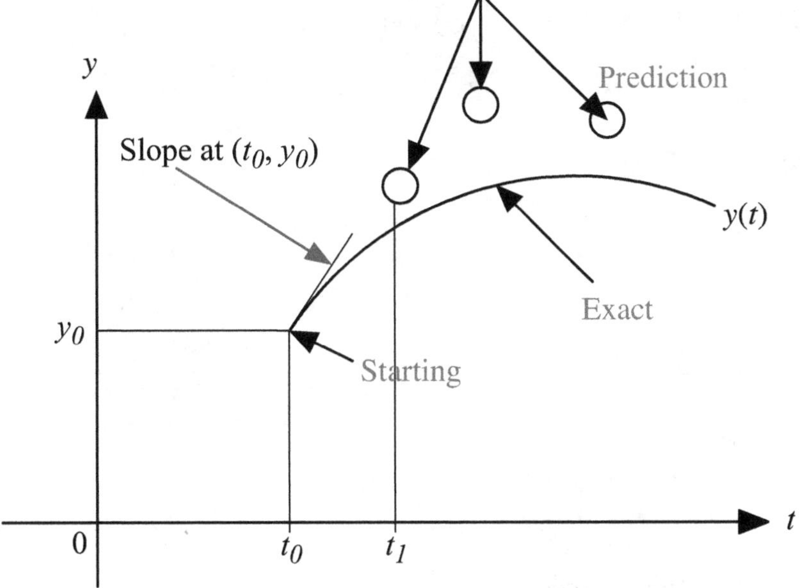

Figure 8.2: Euler's method.

Using the separation of variables, the exact solution is easily obtained as

$$y = e^t.$$

The following C code implements Euler's method for Eq. (8.4):

```
#include <stdio.h>
#include <math.h>
double f(double t, double y)
{
 return y;
}
int main()
{
 double h=0.1, y, t;
 int i;
 t=0.0; y = 1.0;
 printf("t          Euler     Exact\n");
 for (i=0; i<= 10; i++)
  {
```

```
    printf("t= %lf %lf %lf\n", t, y, exp(t));
    y=y+h*f(t,y);
    t=t+h;
   }
  return 0;
}
```

The output is

```
$ gcc euler.c -lm
$ ./a.out
t             Euler     Exact
t= 0.000000 1.000000 1.000000
t= 0.100000 1.100000 1.105171
t= 0.200000 1.210000 1.221403
t= 0.300000 1.331000 1.349859
t= 0.400000 1.464100 1.491825
t= 0.500000 1.610510 1.648721
t= 0.600000 1.771561 1.822119
t= 0.700000 1.948717 2.013753
t= 0.800000 2.143589 2.225541
t= 0.900000 2.357948 2.459603
t= 1.000000 2.593742 2.718282
```

As seen from the table above, the accuracy of Euler's method is marginal at best. Despite its mediocre performance, Euler's method is robust and can be used to obtain a quick result.

Example 2. Lorenz Equations/Strange Attractor/Chaos

Euler's method can be used not only for a single differential equation but for a set of simultaneous differential equations as well. Equations (8.5), known as the Lorenz equations, are an interesting example that demonstrates that a short program in C can produce a surprising result leading to today's research on nonlinear dynamics. Lorenz[1] worked on Eqs. (8.5) that arose in fluid circulation in 1963 which are related to weather patterns. Those equations are nonlinear differential equations which are deterministic yet the behavior of the solution was neither periodic, divergent nor convergent depending on how the parameters, p, R, and b, and the initial

[1]Edward Lorenz (1917–2008) was a mathematician, meteorologist and a pioneer of chaos theory.

values are chosen.

$$\begin{aligned}\frac{du}{dt} &= p(v-u),\\ \frac{dv}{dt} &= -uw + Ru - v,\\ \frac{dw}{dt} &= uv - bw.\end{aligned} \tag{8.5}$$

The following code solves Eqs. (8.5) using Euler's method. The three parameters are chosen as $P = 16.0, b = 4.0$, and $R = 35.0$ and the initial values are chosen as $u = 5.0, v = 5.0, w = 5.0$. The step size is chosen as $h = 0.01$ and 3,000 iterations are performed.

```
#include <stdio.h>
#define P 16.0
#define b 4.0
#define R 35.0

double f1(double u, double v, double w)
{return P*(v-u);}

double f2(double u, double v, double w)
{return -u*w+R*u-v;}

double f3(double u, double v, double w)
{return u*v-b*w;}

int main()
{
double h, t, u, v, w;
int i;

/* initial values */
t=0.0; h=0.01;
u=5.0; v=5.0; w=5.0;

for (i=0; i< 3000; i++)
{
 u=u+h*f1(u,v,w);
 v=v+h*f2(u,v,w);
 w=w+h*f3(u,v,w);
```

```
  printf("%lf %lf\n", u, w);
  t=t+h;
}
  return 0;
}
```

In order to plot (u, w) from the program, gnuplot, a freely available graphic package introduced in Appendix A, can be used. The output from a.out is saved to a data file, lorenz.dat, using I/O redirection and this file is placed or transferred to a directory gnuplot can access to.

```
$ gcc lorenz.c
$ ./a.out  > lorenz.dat
```

Transfer lorenz.dat to a directory accessible by gnuplot. Launch gnuplot and issue the following command:

```
plot 'lorenz.dat' with line
```

The output from gnuplot is shown in Figure 8.3.

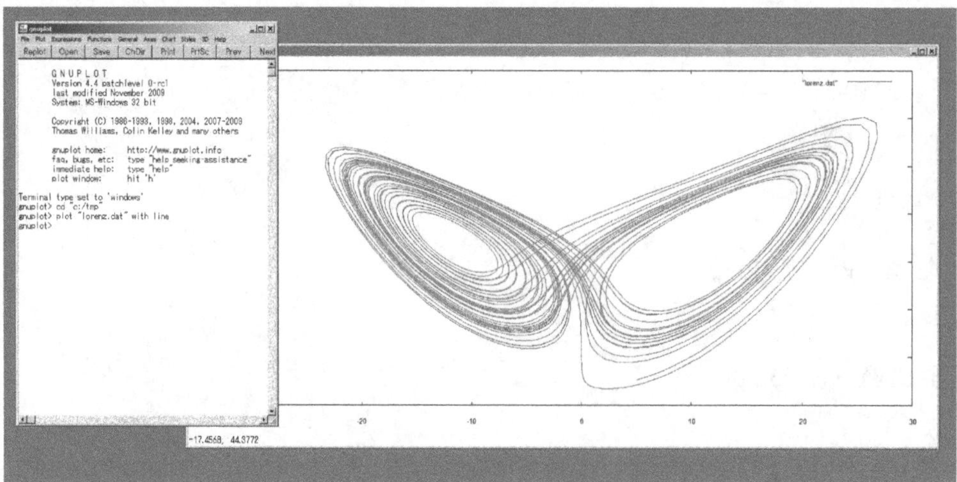

Figure 8.3: Lorenz equations generating chaos.

The pattern in Figure 8.3 was named *chaos*[2] by Lorenz as Eqs. (8.5) are deterministic yet the result of the trajectory, (u, w), in Figure 8.3 is neither periodic, convergent, nor divergent. By

[2]Also called a *strange attractor*.

imposing a small perturbation on the parameters and initial values, the pattern from Eqs. (8.5) changes drastically. Refer to Problem 2 in Exercise 8.3 for another example. As the pattern in Figure 8.3 resembles a butterfly, such observation was later named as the *butterfly effect.*[3]

8.1.2 RUNGE-KUTTA METHOD

The Runge-Kutta[4] method is a refinement of Euler's method in which $f(t, y)$ in Eq. (8.1) is replaced by a weighted average of $k_1 \sim k_4$ defined as

$$\begin{aligned}
k_1 &= f(t, y), \\
k_2 &= f\left(t + \frac{h}{2}, y + \frac{h}{2}k_1\right), \\
k_3 &= f\left(t + \frac{h}{2}, y + \frac{h}{2}k_2\right), \\
k_4 &= f(t + h, y + hk_3).
\end{aligned}$$

Using the above quantities, the iterative scheme is expressed as

$$y_{n+1} = y_n + h\frac{k_1 + 2k_2 + 2k_3 + k_4}{6}. \tag{8.6}$$

The quantities, $k_1 \sim k_4$, are computed at $t, t + \frac{h}{2}$, and $t + h$. The derivation of Eq. (8.6) is more involved and is deferred to an advanced textbook.[5]

The following is a C code to implement the Runge-Kutta method for Eq. (8.4):

```
#include <stdio.h>
#include <math.h>
double f(double t, double y)
{return y;}
int main()
{
 double h=0.1, t, y, k1,k2,k3,k4;
 int i;
/* initial value */
 t=0.0; y=1.0;
 for (i=0; i<=10; i++)
  {
   printf("t= %lf rk= %lf exact=%lf\n", t, y, exp(t));
   k1=h*f(t,y);
```

[3]A movie, The Butterfly Effect (2004), was created after this naming.
[4]Pronounced as *roo ng-uh-koo t-ah*. Both are German mathematicians.
[5]For instance, Iserles, *A First Course in the Numerical Analysis of Differential Equations* (2nd ed.), Cambridge University Press, 2008.

```
    k2=h*f(t+h/2, y+k1/2.0);
    k3=h*f(t+h/2, y+k2/2.0);
    k4=h*f(t+h, y+k3);
    y= y+(k1+2.0*k2+2.0*k3+k4)/6.0;
    t=t+h;
  }
 return 0;
}
```

The output is

```
$ gcc rk4.c -lm
$ ./a.out
t= 0.000000 rk= 1.000000 exact=1.000000
t= 0.100000 rk= 1.105171 exact=1.105171
t= 0.200000 rk= 1.221403 exact=1.221403
t= 0.300000 rk= 1.349858 exact=1.349859
t= 0.400000 rk= 1.491824 exact=1.491825
t= 0.500000 rk= 1.648721 exact=1.648721
t= 0.600000 rk= 1.822118 exact=1.822119
t= 0.700000 rk= 2.013752 exact=2.013753
t= 0.800000 rk= 2.225540 exact=2.225541
t= 0.900000 rk= 2.459601 exact=2.459603
t= 1.000000 rk= 2.718280 exact=2.718282
```

As seen in the output above, the Runge-Kutta method yields much accurate approximation compared with Euler's method. It is the de-fact standard for solving initial value problems.

8.2 HIGHER-ORDER ORDINARY DIFFERENTIAL EQUATIONS

Equation (8.1) is a first-order ordinary differential equation. However, many important and useful differential equations are higher-order differential equations exemplified by the equation of motion (second-order) and the equation of beam deflections (fourth-order).

A higher-order differential equation can be converted to a set of first-order differential equations, thus, any numerical method such as Euler's method or the Runge-Kutta method can be still used to numerically solve for higher-order differential equations.

To illustrate this technique, consider Eq. (8.7), which is the equation for harmonic oscillation that arises in a spring-mass system. It is a second-order ordinary differential equation with two initial conditions:

$$\frac{d^2y}{dt^2} = -y, \quad y(0) = 0, \quad y'(0) = 1. \tag{8.7}$$

Equation (8.7) can be converted into a set of two simultaneous differential equations by setting

$$y_1 \equiv y, \quad y_2 \equiv \frac{dy_1}{dt}. \tag{8.8}$$

With Eq. (8.8), Eq. (8.7) can be rewritten as

$$\frac{dy_1}{dt} = y_2, \tag{8.9}$$

$$\frac{dy_2}{dt} = -y_1, \tag{8.10}$$

with

$$y_1(0) = 0, \quad y_2(0) = 1.$$

The unknowns, $y_1(t)$ and $y_2(t)$, in Eqs. (8.9)–(8.10) can be solved simultaneously. The following code solves for $y1$ and $y2$ using Euler's method:

```
#include <stdio.h>
#include <math.h>
double f1(double x, double y1, double y2)
{return y2;}
double f2(double x, double y1, double y2)
{return -y1;}

int main()
{
 double h=0.01, y1, y2, x;
 int i;
 y1=0.0; y2=1.0;
 x=0.0;
 printf("      x         y2      cos(x)\n");
 for (i=0; i<=10; i++)
 {
  printf("x= %lf %lf %lf\n", x, y2, cos(x));
  y1=y1+h*f1(x,y1,y2);
  y2=y2+h*f2(x,y1,y2);
  x=x+h;
```

```
  }
  return 0;
}
```

In this example, the exact solution for $y2$ is $\cos x$. The output is

```
$ gcc highorder.c -lm
$ ./a.out
        x           y2      cos(x)
x= 0.000000 1.000000 1.000000
x= 0.010000 0.999900 0.999950
x= 0.020000 0.999700 0.999800
x= 0.030000 0.999400 0.999550
x= 0.040000 0.999000 0.999200
x= 0.050000 0.998500 0.998750
x= 0.060000 0.997901 0.998201
x= 0.070000 0.997201 0.997551
x= 0.080000 0.996402 0.996802
x= 0.090000 0.995503 0.995953
x= 0.100000 0.994505 0.995004
```

Similarly, a third-order differential equation and beyond can be also converted into a set of simultaneous first-order differential equations.

8.3 EXERCISES

1. Solve the following differential equation:

$$y' = -x^2 y, \quad y(0) = 1,$$

   by

   (a) exactly (analytically),

   (b) Euler's method, and

   (c) the Runge-Kutta method,

   and plot the three results in a single graph using `gnuplot` (Appendix A). Use $0 \leq x \leq 1$ and $h = 0.1$. Use the following syntax in `gnuplot`:

```
plot 'data1.txt', 'data2.txt', sin(x)
```

where "`data1.txt`" contains data from Euler's method and "`data2.txt`" contains data from the Runge-Kutta method.

2. Numerically solve the following differential equations using Euler's method:

$$\frac{du}{dt} = v,$$
$$\frac{dv}{dt} = -kv - u^3 + B\cos t,$$

with

$$k = 0.1, \quad B = 11.0,$$

and plot the result using `gnuplot`. Change the parameters to

$$k = 0.4, \quad B = 20.0,$$

and show the graph as well. You can use the following as the initial condition:

$$u(0) = 1, \quad v(0) = 1.$$

Also, use

$$h = 0.01, \quad i = 10{,}000.$$

One shows chaos and the other does not.

APPENDIX A

Gnuplot

The C language itself does not support graphics in the standard library as drawing graphics is machine-dependent. To visualize what a C program outputs in C, it is necessary to use machine-specific library files which are not part of the gcc distribution. An alternative is to export data created by a C program to an external application that can read the data file and plot the data.

The graphical application, gnuplot,[1] meets such a requirement. For Windows PCs, download the zipped distribution file from http://www.gnuplot.info. Once downloaded, extract all the files into the same directory and run wgnuplot.exe from there.

The following is a list of the commands in gnuplot that are often used. Comments in gnuplot begins with the "#" symbol. Most of them are self-explanatory. Try typing all by yourself.

```
gnuplot > set title 'My graph' # Prints the title.
gnuplot > plot x**3-x-1 # Note ** (power)
gnuplot > plot sin(x) with dots # w d can be used.
gnuplot > plot sin(x) with impulse # w i also works.
gnuplot > plot [-5:5] sin(x)/(x**2+1) # [-5,5] specifies x range.
# List of functions to plot.
gnuplot > plot [-pi:pi] sin(x), sin(2*x), sin(3*x)
gnuplot > set xlabel 'My x axis'
gnuplot > set ylabel 'My y axis'
gnuplot > plot [-4:4] [0:10] x/exp(x) # Specifies x and y ranges.
gnuplot > splot [-pi:pi] [-2*pi:3*pi] sin(x*y) # splot draws 3-d plot.
```

To draw the graph shown in Figure A.1, enter the following commands:

```
gnuplot> set isosamples 100
gnuplot> set hidden3d
gnuplot> set contour base
gnuplot> splot [0:pi][0:pi] sin(x*y)
gnuplot > quit
```

[1]Freely available for Windows, iOS, and Linux. Source code is also available.

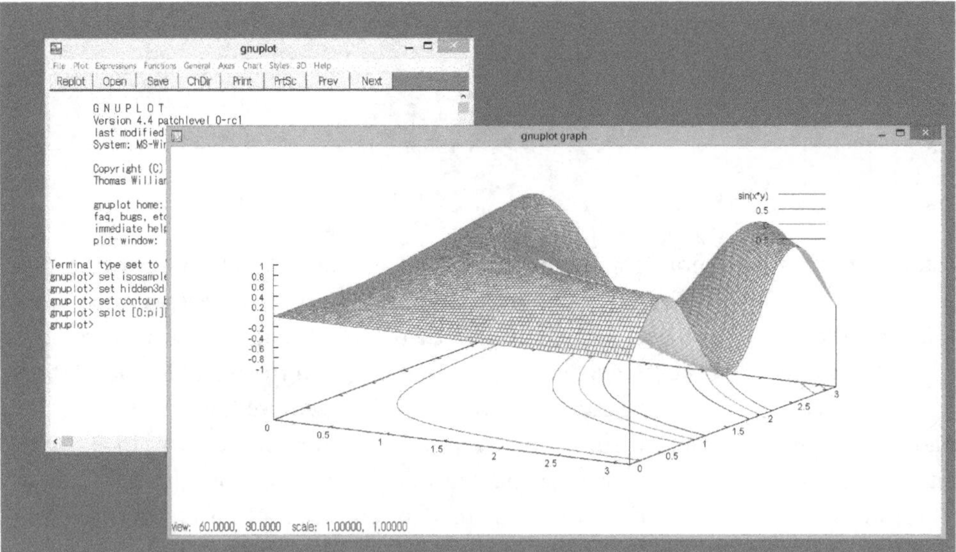

Figure A.1: `gnuplot` example.

Note that two asterisks (**) are used for the power instead of a caret (^). To exit from `gnuplot`, enter `quit`.

If all `gnuplot` can do is to draw a graph of the function entered, it is not very useful. `gnuplot` can draw graphs whose data are created by C. Data generated from a C program can be imported to `gnuplot` for visualization. Here is an example of generating a data file from C and export the file to `gnuplot` to draw the graph.

Prepare the following C code and run with I/O redirection so that the output is stored to a separate file:

```
#include <stdio.h>
#include <math.h>
int main()
{
 int i; float x;
 for (i=0; i<100; i++)
 {
  x = 0.1*i;
  printf("%f %f\n", x , sin(x));
 }
  return 0;
}
```

The output is

```
$ gcc gnuplotdata.c -lm
$ ./a.out > data.dat
```

The output from `a.out` is saved to a file, `data.dat`. Move this file to a directory where `gnuplot` can access. For this purpose, we choose the directory, `C:\tmp` in the Windows system as a working directory.

Now open `gnuplot` and issue the following:

```
gnuplot > cd 'C:\tmp' # Change directory to C:\tmp.
gnuplot > plot 'data.dat' with lines # Plot data in data.dat.
gnuplot > exit
```

Note that the file name, `data.dat`, must be enclosed by the single quotation marks (`'`).[2] Figure A.2 shows the output.

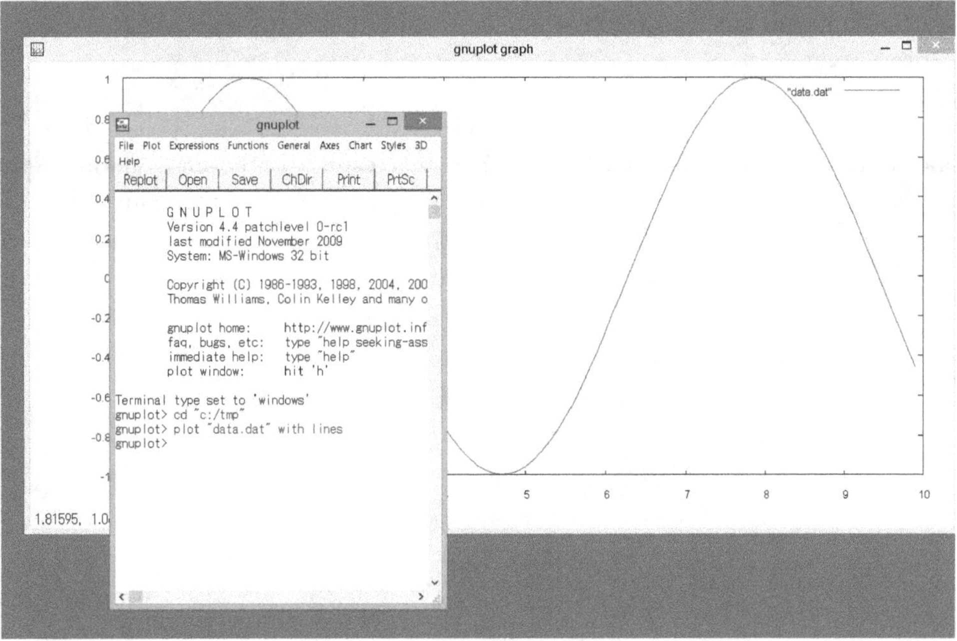

Figure A.2: Plotting a graph from a file.

For another example of plotting data generated by C, refer to Example 2 in Section 8.1.1.

[2]The double quotation mark (`"`) works as well.

To export a graph from gnuplot to another application such as Word, right click the top bar of the graph window, choose Options, and choose Copy to Clipboard. The graphic image is save to memory that can be pasted to any application.

Alternatively, it is possible to save a graph from gnuplot directly to a separate file in the supported graphic formats. The graphic format supported by gnuplot includes jpg, gif, png, and eps. To save a graph in gif format, issue the following:

```
gnuplot > cd 'c:\tmp' # Save file to c:\tmp
gnuplot > set terminal gif size 640, 480 # size is optional
gnuplot > set output 'mygraph.gif'
gnuplot > plot sin(x)
gnuplot > quit
```

To save a graph in eps (enhanced postscript) format, issue the following:

```
gnuplot > cd 'c:\tmp' # Save file to c:\tmp.
gnuplot > set terminal postscript eps enhanced color
gnuplot > set output 'mygraph.eps'
gnuplot > plot sin(x)
gnuplot > quit
```

There are more commands available in gnuplot that cannot be covered in this Appendix. gnuplot also has a capability of being used as a scripting language. Refer to many online tutorials or reference books.[3]

[3]For example, Janert, *Gnuplot in Action: Understanding Data with Graphs*, Manning Publications, 2009.

APPENDIX B

Octave (MATLAB) Tutorial for C Programmers

B.1 INTRODUCTION

MATLAB is a powerful application package for scientific and engineering tasks that combines the easiness of hand-held calculators and the versatility of programming. Many engineering/science classes require students to use MATLAB for homework and projects. This Appendix is a short tutorial of MATLAB for those who already know C but have not learned MATLAB so that they can start working on problems using MATLAB immediately with the least amount of effort. This is possible as many of the MATLAB commands are similar to or variations of corresponding C commands and C programmers can instantly recognize many MATLAB commands without any reference. The converse is not necessarily true, i.e., even with proficiency in MATLAB, mastering C may take a significant amount of time.

There are a few MATLAB look-alikes whose syntax is compatible with MATLAB. All are freely available from the internet including GNU Octave and Scilab. GNU Octave[1] has extensive tools for solving common numerical problems and its syntax is largely compatible with MATLAB. Octave can be run in GUI mode or from a command line as shown in Figure B.1. This Appendix is not meant to be a complete reference of Octave/MATLAB. It is a fast track tutorial for those who learned C and want to pick up the basics of Octave/MATLAB quickly. Numerous reference books on Octave/MATLAB are available on the market.

This Appendix is based on GNU Octave but all the functions and commands in this Appendix are compatible with those of MATLAB.

B.2 BASIC OPERATIONS

B.2.1 PRINCIPLES OF OCTAVE/MATLAB

The following is a list of some of the principles of Octave/MATLAB.

1. There is no distinction between integers and floating points. All variables are double precision by default.

2. Variable names are case-sensitive.

[1]The program can be downloaded from `www.gnu.org/software/octave/`.

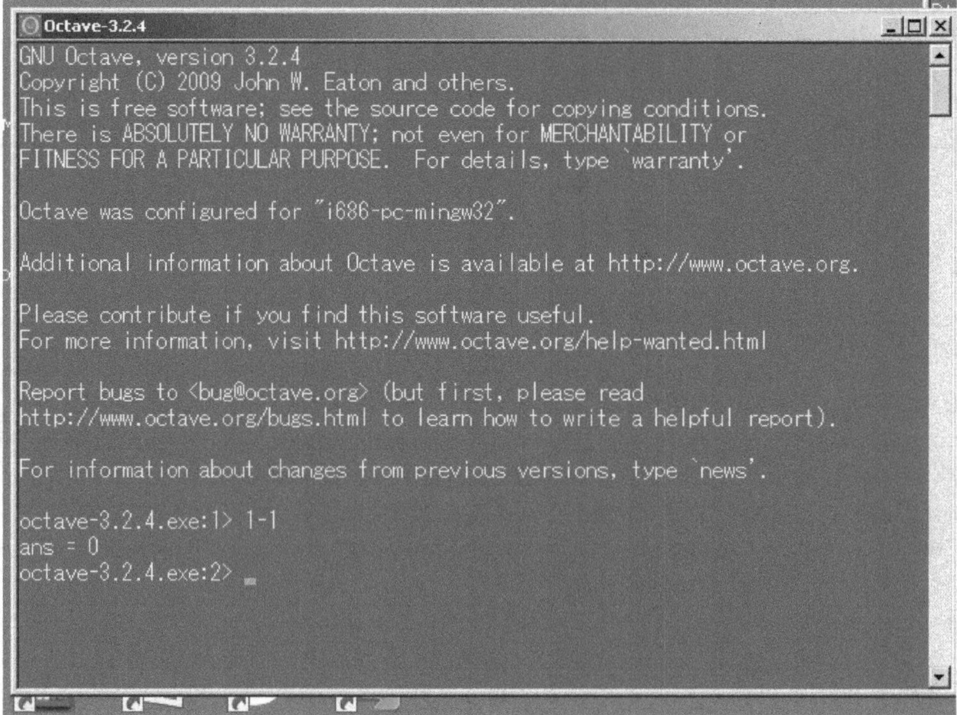

Figure B.1: Octave run from command line.

3. The index of an array begins at 1 (0 for C).
4. Due to its original philosophy,[2] any set of numbers is entered as a matrix including a single variable. An interval is also entered by a matrix.
5. To define vectors and matrices, the square brackets are used, i.e., `[3 4 1]`, `[1 2 3; 4 5 6; 7 8 9]`.
6. Any statement after `%` and `#` is a comment.
7. Statements ending with `;` (semicolon) do not echo back.
8. Both a double quotation mark (`"`) and a single quotation mark (`'`) can be used for a string in Octave but MATLAB only accepts a single quotation mark (`'`).

Try entering the following commands from a command line. Most of the commands are self-explanatory.

[2]MATLAB is an abbreviation for Matrix Laboratory, originally developed for solving problems in linear algebra.

```
clc % Clears screen.
1+3
2^12
(3+2*i)*(2-4*i) % Complex algebra, i*i=-1.
abs(4-3*i) % Absolute value.
sin(pi) % pi is a reserved constant.
cos(2*pi)
log(2.718) % Natural logarithm.
log2(1024) % Logarithm with base 2.
log10(1000) % Logarithm with base 10.
format long % Output in long format.
Just appearance.
sqrt(5)
pi
format short % Output in short format.
Just appearance.
sqrt(3)
```

B.2.2 RESERVED CONSTANTS

Octave/MATLAB has reserved constants. Examples follow:

```
i % Imaginary number, i*i=-1.
j % Same as i, mainly used in electrical engineering.
clock % Current time (six elements).
date % Current date.
pi % 3.14156.
eps % Smallest tolerance number in the system.
```

The reserved constants in Octave/MATLAB can be assigned user supplied values. In the example below, `pi` is assigned a user supplied value of 3.0 and this value is kept until the next `clear pi` command is issued. As a result of this rule, variables such as `i` and `j` can be used as iteration variables even though they were pre-defined as $\sqrt{-1}$.

```
octave.exe:1> pi
ans =  3.1416
octave.exe:2> pi=3.0 % User defined value.
pi =  3
```

```
octave.exe:3> pi*pi % pi=3 is used.
ans =  9
octave.exe:4> clear pi % User defined value cancelled.
octave.exe:5> pi
ans =  3.1416
```

B.2.3 VECTORS/MATRICES

Both vectors and matrices use the square bracket `[ ··· ]` to store their components. To separate each component, use a space or a comma `(,)`. For example,

```
v1=[1 2 3] % Defines a 3-D row vector.
v1=[1, 2, 3] % Same as above.
Separator is either a space or a comma.

v2=[1;2;3] % This defines a column vector.
v2=v1' % Transpose of v1, i.e., column vector.

% Defines a 3x3 matrix.
m1=[1 2 3;4 5 6;7 8 9]
% A semicolon after statement suppresses echo back.
m1=[1,2,3;4,5,6;7,8,9];

m2=[-4 5 6; 1 2 -87; 12 -43 12] % Defines a 3 by 3 matrix.

m2(:, 1) % Extracts the first column.
m2(2, :) % Extracts the second row.

m2(2,3)=10; % Assigns 10 to the (2,3) element of m2.

m1*v2 % Matrix multiplication of m1 times m2.

inv(m1) % Inverse of m1.

[vec, lambda] = eig(m1) % Eigenvectors and eigenvalues of m1.

m1\m2 % Inverse of m1 times m2, same as inv(m1)*m2.
```

```
m1*v2 % Matrix m1 times vector v2.

a=eye(3) % 3x3 identity matrix.

a=zeros(3) % 3x3 matrix with 0 as components.

a=ones(3) % 3x3 matrix with 1 as components.

det(m1) % Determinant of m1.
```

To solve the following three simultaneous equations,

$$\begin{pmatrix} 1 & 4 & 5 \\ 8 & 1 & 2 \\ 6 & 9 & -8 \end{pmatrix} \begin{pmatrix} x \\ y \\ z \end{pmatrix} = \begin{pmatrix} 4 \\ 7 \\ 0 \end{pmatrix},$$

use

```
a=[1 4 5; 8 1 2; 6 9 -8];
b=[4; 7; 0];
sol = inv(a)*b % sol = a\b also works.
```

Although Octave/MATLAB can perform mathematical operations other than linear algebra (matrices/vectors), its underlying design principle is to process numbers as matrices (including vectors). This includes defining an interval range as a row vector and the corresponding function values:

```
octave.exe:1> % A sequence between 1 and 10 with an increment of 2.
octave.exe:1> x=[1: 2 :10]

   1   3   5   7   9

octave.exe:2> %A sequence between 1 and 9 with equidistant 5 entries.
octave.exe:2> x=linspace(1, 9, 5)
x =

   1   3   5   7   9

octave.exe:3> y=x; % Copies x to y.
octave.exe:4> z=x*y % This generates an error message and won't work.
error: operator *: nonconformant arguments (op1 is 1x5, op2 is 1x5)
```

```
octave.exe:4> z=x.*y % This works.
Component wise multiplication.
z =

    1    9   25   49   81

octave.exe:5> z=x./y % Component wise division.
z =

   1   1   1   1   1
```

B.2.4 GRAPH

Both MATLAB and Octave have built-in graphics support. In Octave, built-in `gnuplot` is automatically called for graphics. Try the following graphics commands:

```
x=[0: 0.1 : 10]; % Initial value, increment, final value.
y=sin(x);
plot(x, sin(x)); % Calls gnuplot.

x=linspace(0, 2, 20) % Between 0 and 2 with 20 divisions.
plot(x, sin(x))
%%%%%%%%%%%%%%%%%%
x=[0: 0.2 : 10];
y=1/2 * sin(2*x) ./ x; % ./ divides by individual components.
xlabel('X-axis');
ylabel('sin(2x)/x');
plot(x, y);

close;
%%%%%%%%%%%%%%%%%%
t=[0: 0.02: 2*pi];
plot(cos(3*t), sin(2*t)) % Parametric plot.

%%%%%%%%%%%%%%%%%%%%
x=[0: 0.01: 2*pi];
y1=sin(x);
```

```
y2=sin(2*x);
y3=sin(3*x);
plot(x, y1, x, y2,x, y3); % Plotting multiple graphs.
%%%%%%%%%%%%%
xx=[-10:0.4:10];
yy=xx;
[x,y]=meshgrid(xx,yy);
z=(x .^2+y.^2).*sin(y)./y;
surfc(x,y,z) % 3-D graph.
close
```

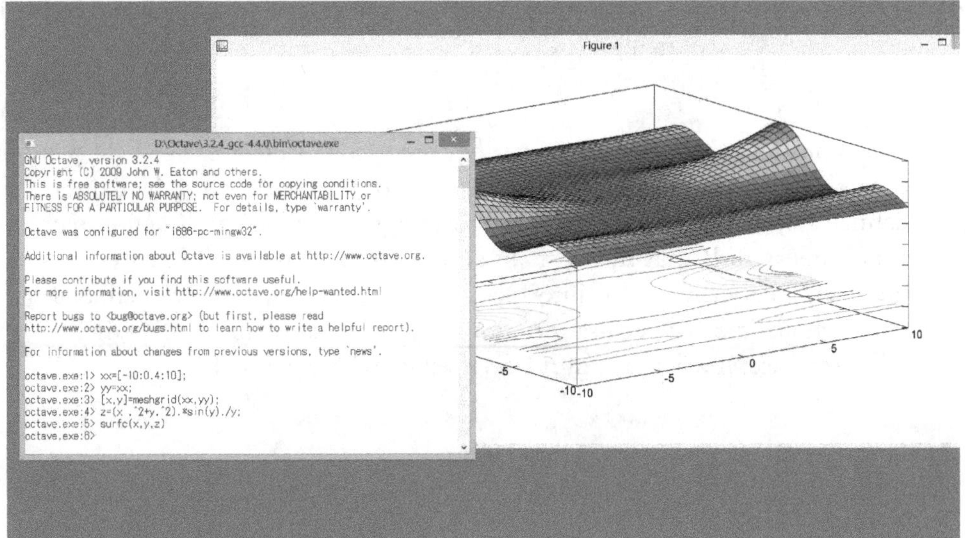

Figure B.2: Graph generated by Octave.

B.2.5 I/O

Octave/MATLAB has several options for input/output. There is no `scanf()` function in Octave/MATLAB. The `input` function can be used as shown in the example below as a combination of `scanf()` and `printf()` in C. The `printf()` function in C is replaced by the `fprintf()` function:

```
% Shows the content of a.
disp(a)
```

```
% Prints a string on screen.
disp('Enter a number = ')
% Prompts for input and value entered is stored in a.
a=input('Enter a number =');
% Same as printf() in C.
fprintf('The solution is %f\n', a);
# %f and %d are available but not %lf in fprintf.
```

B.2.6 M-FILES

There are two types of external files that can be loaded into Octave/MATLAB. They are called *script m-files* and *function m-files*. Both types must have the extension `m` (*.m) and reside in the path Octave/MATLAB can search.

1. Function m-files

   In C, functions must be declared before they are used. In Octave/MATLAB, a separate file has to be prepared to use a user-defined function. A file that has the definition of the user-defined function must be saved under the same name as the function name with `m` as the file extension. If that file exists in a directory Octave/MATLAB can access, that function is automatically loaded into Octave/MATLAB and can be used as if it were a built-in function. It is not necessary to explicitly load that m-file (in fact, it will cause an error). For instance, if a file, `myfunction.m`, exists whose content is

   ```
   function y=myfunction(x)
   y=x^3-x+1;
   ```

   the function, `myfunction(x)`, is automatically available in Octave/MATLAB. A function can return multiple values. In the following example, when `z` is input, `x` stores `z^2` and `y` stores `z^3`.

   ```
   function [x, y]=myfunction2(z)
   x=z^2;
   y=z^3;
   ```

   In Octave/MATLAB, this function can be called as

   ```
   [a,b]=myfunction2(3);
   ```

There is a way to define a function without a separate file using *anonymous functions*. Use the following example:

```
octave.exe:20> f=@(x,y) x-y^2;
octave.exe:21> f(1,2)
ans = -3
```

2. Script m-files

   A script m-file is a batch file that contains a set of statements that are otherwise typed from the keyboard. Any Octave/MATLAB statement can be included in a script m-file except for function definitions that must be separately saved as function m-files. To load a script m-file, type the file name without extension, `m`.[3] For instance, consider a file whose name is `myscript.m` prepared with the following content saved in the `C:\tmp` directory:

```
a=input('Enter a number=');
fprintf('The square of %f is %f\n.', a, a^2);
```

   In an Octave/MATLAB session, this script m-file can be executed as

```
octave.exe:10> cd 'C:\tmp' % Changes the working directory.
octave.exe:11> myscript % Note no extension (m).
Enter a number=12
The square of 12.000000 is 144.000000
```

B.2.7 CONDITIONAL STATEMENT

The conditional statements in Octave/MATLAB are similar to those in C with slight modification of the syntax. The following examples are self-explanatory:

1. If statement

   The following example shows the usage of `if` statements. Note that `if` must be paired with end:

```
if a>2
 disp('a is larger than 2.')
 else
```

[3]A script m-file name should not contain a minus (-) sign. Why?

```
 disp('a is smaller than 2.')
end
```

2. For statement

   The following example shows the usage of `for` statements. Note that `for` must be paired with end:

```
for k=0:2:10 % k goes from 0 to 10 with an increment of 2.
 disp(2*k)
end
```

B.3 SKETCH OF COMPARISON BETWEEN C AND OCTAVE/MATLAB

For C programmers, the best way to quickly pick up the syntax of Octave/MATLAB is to compare two programs written in each language side by side. The following shows a side-by-side comparison of programs that use the same logic written in C and Octave/MATLAB.

1. Programs to solve quadratic equations:

```c
/* This program computes two
   roots for a quadratic
   equation. */

#include <stdio.h>
#include <math.h>

int main()
{
double a, b, c, disc, x1, x2;

printf("Enter 3 coeffs =");
scanf("%lf %lf %lf", &a, &b, &c);

disc=b*b-4*a*c;
if (disc<0)
  {printf("Imaginary roots !\n");
   return 0;}
x1= (-b + sqrt(disc))/(2*a);
x2= (-b - sqrt(disc))/(2*a);
printf("The roots are "
   "%lf, %lf.\n", x1, x2);
return 0;
}
```

```matlab
% This program computes two
%      roots
% for a quadratic equation.

a=input("Enter a = ");
b=input("Enter b = ");
c=input("Enter c = ");

disc=b^2-4*a*c;

if disc<0
 disp('Imaginary roots !');
 return;
end;

x1=(-b-sqrt(disc))/(2*a);
x2=(-b+sqrt(disc))/(2*a);

fprintf('Roots are %f %f.\n',
         x1,x2);
```

2. Programs for series summation:

```
#include <stdio.h>
#include <math.h>
int main()
{
  int i;
  double sum = 0.0;
  for (i=0; i<1000; i++)
    sum+=pow(-1,i)/(double)(2*i+1);
  printf("approx= %lf
    exact value= %lf.\n",
   4*sum, 4*atan(1.0));
return 0;
}
```

```
sum=0;

for k=0:1000
 sum=sum+(-1)^k/(2*k+1);
end

fprintf('Approx and exact
 values = %f %f.\n',
    4*sum, 4*atan(1));
```

Note that sum is a reserved function in Octave/MATLAB.

3. Programs to handle arrays:

```
#include <math.h>
#define N 10
int main()
{
  float x[]={ -4.0, 1.2, 1.3, 2.5,
              -12.7, 9.0,1.41,
  65.2, -2.1, 2.36};
  int i;
  float sum = 0.0, average,
        variance;
  for (i=0; i<N; i++) sum+=x[i];
  average=sum/N;

  for (i=0; i<N; i++)
   variance += pow(x[i] -
              average, 2);
  variance = variance/(N-1);

  printf("avg.= %f std. dev. ="
         "%f \n ", average,
 sqrt(variance));
  return 0;
}
```

```
N=10;
x=[-4 1.2 1.3 2.5 -12.7 9 1.41
 65.2 -2.1 2.36];
sum=0;
for k=1:N
 sum=sum+x(k)
end
ave=sum/N;
var=0;
for k=1:N
 var=var+(x(k)-ave)^2
end
var=var/(N-1);
fprintf('Average = %f,
 Standard deviation= %f.\n',
  ave, sqrt(var));
```

4. Programs to use functions:

```
#include <stdio.h>
#include <math.h>
#define EPS 1.0e-6
double f(double x)
 {return x*x-2;}

double fp(double x)
 {return 2*x;}
double newton(double x)
{ return x - f(x)/fp(x);}
int main()
{
 double x1, x2;
 int i;
 printf("Enter initial guess  =");
 scanf("%lf", &x1);
 if (fp(x1)==0.0) {
 printf("No convergence.\n");
 return 1; }
 for (i=0;i<100;i++)
 {
  x2=newton(x1);
  if (fabs(x1-x2)< EPS) break;
  x1=x2;
 }

 printf("iteration = %d\n", i);
 printf("x= %lf\n", x1);
 return 0;
}
```

```
function y=f(x)
y=x*x-2;
%
% Save this file as f.m
%
function y=fp(x)
 y=2*x;
%
% Save this file as fp.m
%
function y=newton(x)
 y=x-f(x)/fp(x);
%
% Save this file as newton.m
%
x1=input('Enter initial
 guess = ');

if fp(x1)<eps
 disp('No convergence !');
 return; end

for i=0:1:99
 x2=newton(x1);
 if abs(x1-x2)<1e-10 break;end
 x1=x2;
end

fprintf('Iteration = %d\n', i);
fprintf('x = %f\n', x1);
```

5. Programs to handle external files:

```
#include <stdio.h>
int main()
{
   FILE *fp1, *fp2;
   float a,b,c;
   fp1=fopen("data1.dat","r");
   fscanf(fp1, "%f", &a);
   fclose(fp1)

   fp2=fopen("data2.dat","w");
   fprintf(fp2,"This is the "
    "first file.\n");
   fclose(fp2);

   return 0;
}
```

```
fp1=fopen('data1.dat', 'r');
a=fscanf(fp1, '%f');
fclose(fp1);

fp2=fopen('data2.dat', 'w');
fprintf(fp2, 'This is the
  first line.\n');
fclose(fp2);
```

B.4 EXERCISES

1. Translate the following C code to an equivalent Octave/MATLAB m-file:

```
#include <stdio.h>
int main()
{
 double x, y, z;
 int i,n;
 x=y=z=0.0;
 printf("Enter # of iteration = ");
 scanf("%d", &n);
 for (i=0; i<n; i++)
  {
   x = (10-y-2*z)/7;
   y = (8-x-3*z)/8.0;
   z = (6-2*x-3*y)/9.0;
  }
 printf("x = %lf, y= %lf, z=%lf.\n", x,y,z);
```

```
 return 0;
}
```

Note the following syntax:

```
fprintf('x = %f, y= %f, z=%f\n', x,y,z);
```

2. Translate the following C code to equivalent Octave/MATLAB m-files. You need to prepare two m-files, one for a function m-file (`f.m`) and the other for a script m-file (`simpson.m`):

```
/* Simpson's rule */
#include <stdio.h>
#include <math.h>

double f(double x)
 {return 4.0/(1.0+x*x);}
int main()
{
 int i, n ;
 double a=0.0, b=1.0 , h, s1=0.0, s2=0.0, s3=0.0, x;
 printf("Enter number of partitions (must be even) = ");
 scanf("%d", &n) ;
 h = (b-a)/(2.0*n) ;
 s1 = (f(a)+ f(b));
 for (i=1; i<2*n; i=i+2) s2 = s2 + f(a + i*h);
 for (i=2; i<2*n; i=i+2) s3 = s3 + f(a + i*h);
 printf("%lf\n", (h/3.0)*(s1+ 4.0*s2 + 2.0*s3));
 return 0;
}
```

APPENDIX C

FORTRAN Tutorial for C Programmers

FORTRAN (FORmula TRANslation) was created in 1957 at IBM and was the de-facto programming language for engineers and scientists until around the 1990s because of the vast amount of subroutines written (and fully debugged) in FORTRAN. Today, FORTRAN is still the major computational language for some areas in science/engineering (computational fluid mechanics for example) and FORTRAN is used as the benchmark for supercomputers. FORTRAN has been evolved since its inception to adopt modern concepts in computer languages and FORTRAN 2018 is available as of this writing.

In this Appendix, the focus is for C programmers to be able to at least read legacy codes written in FORTRAN 77 quickly and possibly make a minor modification. A plausible situation would be that a FORTRAN code written 30 years ago is given to a C programmer who needs to modify part of the code but is not necessarily required to write a code in FORTRAN from scratch. The C programmer needs to know only the basic syntax of FORTRAN which can be inferred from the C syntax.

With this Appendix, the C programmer should be able to achieve this goal. This Appendix is not meant to be a complete reference to the FORTRAN language, as there are many good reference books available on the market.

C.1 FORTRAN FEATURES

As FORTRAN programs were prepared using a deck of IBM punch cards shown in Figure C.1 one line per card, the width of the card (80 characters) was all it was able to utilize. Unlike modern computer languages, FORTRAN has the following restrictions:

1. Case insensitive. Originally only uppercase letters were used but today a program can mix both uppercase and lowercase letters.

2. Not free-form, The entire line is limited to 80 columns wide.

   (a) Column 1 is reserved for comments. If any character is written in Column 1, that line is interpreted as a comment line.

   (b) Columns 2–5 are reserved for line identification (mainly used by `GOTO` statements).

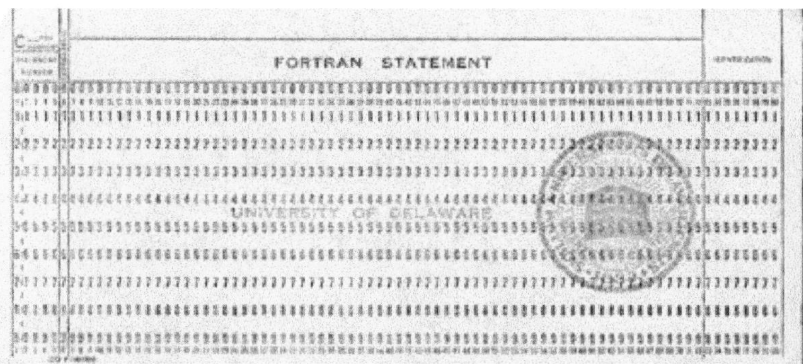

Figure C.1: IBM punch card.

   (c) Column 6 is reserved for continuation. If any character is written on Column 6, that line is interpreted as a continuation line from the previous line.

   (d) Columns 7–72 are reserved for a FORTRAN statement. This is the only space a FORTRAN code can be written to.

3. Variables beginning with I-N are automatically declared as integers by default. All other variables are implicitly declared as real numbers. To override this rule, it is necessary to explicitly declare variables as INTEGER, REAL, or DOUBLE PRECISION.

4. No mixture of different types of variables is allowed. If an integer variable needs to be used that is evaluated to be a floating number, use the `FLOAT` function, i.e., `FLOAT(I)`. Similarly, for an integer constant, change `3` to `3.0`, for example, to be mixed with floating variables.

C.2 HOW TO RUN A FORTRAN PROGRAM

As FORTRAN is a compiled language, it is run much the same way as C programs are run.

1. UNIX system

   On a typical UNIX system, either `g77` (GNU FORTRAN compiler) or `f77` (hardware/software manufacturer supplied FORTRAN compiler) is available. Instead of running `gcc`, run `g77` followed by the name of FORTRAN file. The file extension of FORTRAN file must be `.f` or `.f77`.

```
$ nano MyProgram.f
$ g77 MyProgram.f
$ ./a.out
```

2. Windows system

   Free FORTRAN, `g77`, for the Windows system is available from the same site that `gcc` was downloaded. As the site to download changes over the time, do Google search for keywords such as `gnu g77 windows` for the current download site.

   To compile a FORTRAN program, issue

```
g77 MyProgram.c
```

   This will produce an executable file, `a.exe` instead of `a.out`.

C.3 SKETCH OF COMPARISON BETWEEN C AND FORTRAN

Just like Octave/MATLAB, the best way to pick up the FORTRAN syntax for a C programmer is to compare programs written in C and FORTRAN side by side. The following lists the same programs used in the Octave/MATLAB section but with the corresponding FORTRAN programs.

1. Programs to solve quadratic equations:

```
/* This program computes roots
for quadratic equation. */
#include <stdio.h>
#include <math.h>

int main()
{
double a, b, c, disc, x1, x2;

printf("Enter 3 coeffs =");
scanf("%lf %lf %lf", &a, &b, &c);

disc=b*b-4*a*c;

if (disc<= 0)
  {printf("Imaginary roots !\n");
   return 0;}

x1= (-b + sqrt(disc))/(2*a);
x2= (-b - sqrt(disc))/(2*a);

printf("The roots are %lf and
  %lf. \n", x1, x2);
return 0;
}
```

```
C     This program computes roots
C        for quadratic equations.
      DOUBLE PRECISION A, B, C,
     c  DISC, X1,X2
      WRITE(*,*)
     c "Enter three coeffs"
      READ(*,*) A, B, C
      DISC=B**2-4.0*A*C

      IF(DISC.LE.0.0) THEN
       WRITE(*,*)
     c  "Imaginary roots."
      STOP
      ENDIF

      X1=(-B+SQRT(DISC))/(2.0*A)
      X2=(-B-SQRT(DISC))/(2.0*A)
      WRITE(*,*) "Roots are ",
     c  X1, X2
      STOP
      END
```

Note:

- The `WRITE(*,*) "STRING", A` line means to write `STRING` and the value of `A` to the default device (the first `*`, screen) using the default format (the second `*`).
- `A` to the power of `B` (A^B) can be entered as `A**B`.
- The following is a list of relational operators used in `IF`:
  - `A.LE.B` A is less than or equal to B (`a<= b`).
  - `A.LT.B` A is less than B (`a < b`).
  - `A.EQ.B` A is equal to B (`a == b`).
  - `A.GT.B` A is greater than B (`a > b`).
  - `A.GE.B` A is greater than or equal to B (`a >= b`).
  - `A.NE.B` A is not equal to B (`a != b`).

- `A.AND.B` A and B (`a && b`).
- `A.OR.B` A or B (`a || b`).

2. Programs for series summation:

```
#include <stdio.h>
#include <math.h>

int main()
{
  int i;
  double sum = 0.0;
  for (i=0; i<1000; i++)
    sum+=pow(-1,i)/
         (double)(2*i+1);
  printf("approx= %f true value= "
    "%lf\n ", 4*sum, 4*atan(1.0));
return 0;
}
```

```
      DOUBLE PRECISION SUM
      INTEGER I
      SUM=0.0

      DO 10 I=0,999,1
       SUM=SUM+(-1)**I/
     c  (2.0*FLOAT(I)+1.0)
 10   CONTINUE

      WRITE(*,*) "Approx and
     c  exact values =", 4.0*SUM,
     c  4.0*ATAN(1.0)
      STOP
      END
```

- There is no `FOR` statement available in FORTRAN. For iterations, `DO` and `CONTINUE` are used.
- `DO 10 I=0,999,1` means to repeat the statements between the `DO` line and the line that has the line number `10` with the iteration variable, `I`, from 0 to 999 with the increment of 1. The CONTINUE statement is to literally keep going doing nothing else.

3. Programs to handle arrays:

```c
#include <stdio.h>
#include <math.h>
#define N 10
int main()
{
  float x[]={ -4.0, 1.2, 1.3, 2.5,
   -12.7, 9.0, 1.41, 65.2, -2.1,
    2.36};
  int i;
  float sum = 0.0, average,
        variance;

  for (i=0; i<N; i++) sum+=x[i];
  average=sum/N;

  for (i=0; i<N; i++)
   variance += pow(x[i] -
               average, 2);
  variance = variance/(N-1);

  printf("avg.= %f std. "
         "dev. = %f \n ",
    average, sqrt(variance));
  return 0;
}
```

```fortran
      PARAMETER(N=10)
      REAL X(N), SUM, AVE, VAR
      INTEGER I

      DATA X /-4.0,1.2,1.3,2.5,
     c -12.7,9.0,1.41,
     c 65.2,-2.1, 2.36/

      SUM=0.0

      DO 10 I=1,N
       SUM=SUM+X(I)
 10   CONTINUE

      AVE=SUM/FLOAT(N)

      DO 20 I=1,N
       VAR=VAR+(X(I)-AVE)**2
 20   CONTINUE
      VAR=VAR/(FLOAT(N)-1.0)

      WRITE(*,*) "Average=",
     c  AVE,
     c  ' Standard deviation =',
     c  SQRT(VAR)
      STOP
      END
```

- `PARAMETER` can specify a constant.
- `FLOAT` converts an integer variable to a floating number.

4. Programs to use functions:

```
#include <stdio.h>
#include <math.h>
#define EPS 1.0e-6

double f(double x)
{
return x*x-2;
}

double fp(double x)
{
return 2*x;
}

double newton(double x)
{
return x - f(x)/fp(x);
}

int main()
{
double x1, x2;
int i;

printf("Enter initial guess  =");
scanf("%lf", &x1);
```

```
******************************
      FUNCTION F(X)
      DOUBLE PRECISION F, X
      F=X*X-2.0
      RETURN
      END
******************************
      FUNCTION FP(X)
      DOUBLE PRECISION FP, X
      FP=2.0*X
      RETURN
      END
******************************
      FUNCTION NEWTON(X)
      DOUBLE PRECISION NEWTON,
     c  X, F, FP
      NEWTON=X-F(X)/FP(X)
      RETURN
      END
****************************

      PARAMETER(EPS=1.0E-6)
      DOUBLE PRECISION X1, X2,
     c  NEWTON, F, FP
      INTEGER I

      WRITE(*,*)
     c  "Enter initial guess"
      READ(*,*) X1
```

```
if (fp(x1)==0.0) {                         IF(FP(X1).EQ.0.0) THEN
printf("No convergence.\n");                 WRITE(*,*) "No conv"
return;                                      STOP
}                                          ENDIF

for (i=0;i<100;i++)                        DO 10 I=0,99,1
{                                           X2=NEWTON(X1)
x2=newton(x1);                              IF(ABS(X1-X2).LT.EPS)
if (fabs(x1-x2)< EPS) break;          c        GOTO 11
x1=x2;                                      X1=X2
}                                     10   CONTINUE

printf("iteration = %d\n", i);        11   CONTINUE
printf("x= %lf\n", x1);
return 0;                                  WRITE(*,*)"Iteration =",I
}                                          WRITE(*,*)"X=",X1

                                           STOP
                                           END
```

- The type of of each function defined must be also declared within the main program.

5. Programs to handle external files:

```
#include <stdio.h>
int main()
{
    FILE *fp1, *fp2;
    float a,b,c;
    fp1=fopen("data11.dat","r");
    fscanf(fp1, "%f", &a);
    fclose(fp1)

    fp2=fopen("data2.dat","w");
    fprintf(fp2,"This is the "
                "first file.\n");
    fclose(fp2);

    return 0;
}
```

```
      REAL A, B, C
      B=3.14
C
      OPEN(UNIT=1,
     c  FILE='data1.dat',
     c  STATUS='old')
      READ(1, *) A
      WRITE(*,*) A
      CLOSE (1)
C
      OPEN(UNIT=2,
     c  FILE='data2.dat',
     c  STATUS='new')
      WRITE(2,*) 'This is
     c  the first file.'
      WRITE(2,*) B
      CLOSE(2)
C
      STOP
      END
```

- To access an external file, use `OPEN` to open the file, assign a unit number to `UNIT` and specify whether the file is for writing (`STATUS='new'`) or for reading (`STATUS='old'`).
- `READ(1, *) A` means to read a variable, `A`, from a unit 1 (in this case, an external file, `data1.dat`) using default format.
- `WRITE(2, *) B` means to write the value of `B` to a unit 2 (in this case, an external file, `data2.dat`) using default format.

C.4 EXERCISES

1. Translate the following C code to a FORTRAN code:

```
#include <stdio.h>
#define N 10
int main()
{
 float x[N]={1, 2, 3, 4, 5, 6, 7, 8, 9, 10},
 y[N]={549.88, 693.932, 415.337, 624.482,
  436.095, 355.256, 185.603,
  308.003, 244.414, 376.182};
 float xysum=0.0, xsum=0.0, ysum=0.0, x2sum=0.0;
 float a, b;
 int i;

 for (i=0; i<N ; i++)
  {
    xsum+=x[i];
    ysum+=y[i];
    xysum+=x[i]*y[i];
    x2sum+=x[i]*x[i];
  }
 a = (xysum*N-xsum*ysum)/(x2sum*N-xsum*xsum);
 b = (x2sum*ysum-xysum*xsum)/(x2sum*N-xsum*xsum);
 printf("%f %f \n", a, b);
 return 0;
}
```

2. Translate the following C code to a FORTRAN code:

```
#include <stdio.h>
#include <math.h>
double f(double t, double y)
{return y;}

int main()
{
 double h=0.1, t, y, k1,k2,k3,k4;
```

```
 int i;

/* initial value */
 t=0.0; y=1.0;

 for (i=0; i<=10; i++)
  {
    printf("t= %lf rk= %lf exact=%lf\n", t, y, exp(t));
    k1=h*f(t,y);
    k2=h*f(t+h/2, y+k1/2.0);
    k3=h*f(t+h/2, y+k2/2.0);
    k4=h*f(t+h, y+k3);
    y= y+(k1+2.0*k2+2.0*k3+k4)/6.0;
    t=t+h;
  }
 return 0;
}
```

Author's Biography

SEIICHI NOMURA

Seiichi Nomura is a Professor in the Department of Mechanical and Aerospace Engineering at the University of Texas at Arlington. He is the author of *Micromechanics with Mathematica* and coauthor of *Heat Conduction in Composite Materials* with A. Haji-Sheikh. His research interests include micromechanics, analysis of composite materials, and applications of computer algebra systems. He holds a Dr. of Eng. degree from the University of Tokyo and a Ph.D. from the University of Delaware.

Index